Garnets: Their Mining, Milling and Utilization

by US Bureau of Mines

with an introduction by Kerby Jackson

Introduction

It has been years since the U.S. Bureau of Mines released their important publication "Garnet: Its Mining, Milling and Utilization". First released in 1925, this work has been unavailable to the mining community since those days, with the exception of expensive original collector's copies and poorly produced digital editions.

It has often been said that "*gold is where you find it*", but even beginning prospectors understand that their chances for finding something of value in the earth or in the streams of the Golden West are dramatically increased by going back to those places where gold and other minerals were once mined by our forerunners. Despite this, much of the contemporary information on local mining history that is currently available is mostly a result of mere local folklore and persistent rumors of major strikes, the details and facts of which, have long been distorted. Long gone are the old timers and with them, the days of first hand knowledge of the mines of the area and how they operated. Also long gone are most of their notes, their assay reports, their mine maps and personal scrapbooks, along with most of the surveys and reports that were performed for them by private and government geologists. Even published books such as this one are often retired to the local landfill or backyard burn pile by the descendents of those old timers and disappear at an alarming rate. Despite the fact that we live in the so-called "Information Age" where information is supposedly only the push of a button on a keyboard away, true insight into mining properties remains illusive and hard to come by, even to those of us who seek out this sort of information as if our lives depend upon it. Without this type of information readily available to the average independent miner, there is little hope that our metal mining industry will ever recover.

Though this volume may not at first seem to be of great importance to gold miners, I feel that those miners with an interest in smelting and refining their finds, especially those recovered from lodes, will find the processes outlined to be of great value.

This important volume and others like it, are being presented in their entirety again, in the hope that the average prospector will no longer stumble through the overgrown hills and the tailing strewn creeks without being well informed enough to have a chance to succeed at his ventures.

Please note that at times it is necessary to rearrange illustration plates in these texts. Any illustrations not found in their original sequence may be found following the index.

Kerby Jackson
Josephine County, Oregon
June 2015

www.goldminingbooks.com

GARNET: ITS MINING, MILLING, AND UTILIZATION

By W. M. MYERS and C. O. ANDERSON

INTRODUCTION

The term "garnet" is popularly associated with a dark-red, semi-precious stone that has been used for ornament since prehistoric times. Only within recent years has it been recognized that an unusual combination of physical properties makes this mineral, or, rather, group of minerals, useful to modern industry. The recognition of the abrasive qualities of garnet and the superiority of garnet as an abrasive for some purposes resulted in the active search for deposits that could be worked at a profit on a commercial scale. Such deposits have been found and are being worked by up-to-date methods; the problem of separating the garnet from its associated minerals has been solved; and the production of garnet has become a well-established industry. This bulletin presents the results of an investigation conducted by the Bureau of Mines as part of its work for the increase of efficiency and the prevention of waste in the mineral industries.

ACKNOWLEDGMENTS

The authors wish to acknowledge the hearty cooperation of all the garnet producers mentioned in this bulletin, who freely gave permission to examine their properties and furnished a great deal of technical data. Particular acknowledgment is made of the many courtesies extended by Mr. C. R. Barton, Mr. T. S. Mennie, Mr. F. C. Hooper, Mr. N. Davenport, and Mr. J. Davenport. For information concerning the use and manufacture of garnet products thanks are due to the H. H. Barton & Son Co., Herman Behr & Co. (Inc.), the American Glue Co., the Wausau Abrasive Co., the United States Sand Paper Co., the Manning Abrasive Co., and the American Gem and Pearl Co.

MINERALOGY OF GARNET

The term "garnet" is applied to a closely related group of minerals which crystallize in the same forms and have similar physical properties, although their chemical composition may vary widely.

The term "garnet" is derived from the Latin granatus, seedlike, in allusion to the appearance of the crystals embedded in their matrix. The early English term "granat" was derived directly from the Latin and it is only since the eighteenth century that the present term "garnet" has come into common use.

CRYSTALLOGRAPHY

Garnet crystallizes in the cubic system—commonly in rhombic dodecahedrons, in tetragonal trisoctahedrons, or in combinations of these forms. The hexoctahedron appears occasionally; other crystal forms are rare. Under favorable conditions garnet crystallizes in remarkably perfect forms which exhibit crystal faces rivaling the facets of a cut gem. Under less favorable conditions garnets appear either in irregular masses on which crystal faces can just be distinguished or in rounded grains disseminated throughout the rock. Occasionally the garnet is so segregated from the associated minerals that it forms large massive aggregates.

PHYSICAL PROPERTIES

The physical properties of the garnet group vary with the individual varieties, but in general fall within the following limits: Color, widely variant, from colorless and white through all shades of yellow, brown, red, and green to black; transparency, transparent to opaque; luster, vitreous, resinous, or dull; streak, white; specific gravity, 3.4 to 4.3.

Fusibility.—The fusibility is variable and depends on chemical composition. Garnets containing considerable iron fuse readily before the blowpipe to a dark glass, at temperatures near 1,300° C. Garnet containing a considerable percentage of chromium is infusible before the blowpipe. When fused by itself, garnet disassociates into other compounds. In the fused mass a number of minerals have been identified, bearing no relation to garnet other than that their chemical constitutents were derived from it.

Index of refraction.—The index of refraction of the garnet group ranges from 1.735 to 1.94. As garnet crystallizes in the isometric or cubic system it has only one index. Some garnets, however, occasionally display an anomalous double refraction which is believed to be due to a complex twining of triclinic individuals that produces forms apparently isometric.

Cleavage.—An indistinct dodecahedral cleavage has been observed rarely. Some species of garnet have a pronounced laminated structure which causes planes of weakness along which the garnet separates. This parting is mechanical, is not related to the crystal form, and hence can not be considered a true cleavage.

Fracture.—The fracture of garnet shows great variation. In some varieties, particularly those having a glassy structure, it is decidedly conchoidal, for the mineral tends to break in thin, shell-like flakes. Other varieties show the conchoidal fracture less plainly or not at all and their fracture may be termed uneven.

Tenacity.—Aggregates of garnet composed of many small individuals are brittle and shatter readily; massive garnet and well-formed crystals are remarkably tough and are shattered with difficulty.

Hardness.—The hardness of garnet ranges from 6.5 to 7.5 in Moh's scale. Some specimens are said to have had a hardness approaching 8.0. Sound crystallized garnet generally has a hardness of 7.5; it is therefore slightly harder than quartz, which has a hardness of 7.0.

CHEMICAL COMPOSITION

The chemical composition of the garnet group may be represented by the general formula $R''_3R'''_2.3SiO_4$ or $3R''O.R'''_2O_3.3SiO_2$, in which R'' represents the bivalent elements calcium, magnesium, manganese in the manganous state, and ferrous iron, and R''' represents the trivalent elements, ferric iron, aluminum, and chromic chromium. Rarely a part of the ferric iron or silicon is replaced by titanium. Because of the isomorphous substitution of the different bivalent and trivalent elements, specimens of garnet with a constitution that may be represented by a definite formula are rare and the composition of any given specimen is generally very complex.

The chemical stability of garnet varies greatly with composition; some garnets have been exposed on the surface of a rock outcrop for long periods of time—probably since the glacial age—without undergoing alteration. Other garnets have been completely altered and pseudomorphs after garnet are common; in these, although the original crystal form remains unchanged, the garnet has altered to scapolite, epidote, oligoclase, hornblende, chlorite, or other minerals. Garnets containing ferrous iron may disintegrate through the oxidation of the iron, which generally forms a rusty coating of limonite around fragments of unaltered garnet. Members of the garnet group have a marked tendency to include other minerals within their crystals. In some specimens this tendency is so strong that the garnet crystal is only a shell inclosing other minerals. The included minerals often are quartz, mica, or pyroxene, which are so finely disseminated that their mechanical separation from the inclosing garnet is very difficult, and to obtain samples of pure garnet for analysis is troublesome.

VARIETIES OF GARNET

The garnet group is composed of six species and their isomorphous mixtures. One of these mixtures, rhodolite, which is said to exist in fairly definite proportions, may be said to constitute a seventh species. The garnet group may be further subdivided into three smaller groups in which the species of garnet are classified according to the preponderance of the trivalent elements, aluminum, ferric iron, and chromic chromium. The aluminum garnets are grossularite, pyrope, almandite, rhodolite, and spessartite. The ferric iron variety of garnet is represented by andradite and the chromium variety by uvarovite.

Grossularite.—Grossularite, also known as hessonite, essonite, or cinnamon stone, is represented by the formula $Ca_3Al_2(SiO_4)_3$. Its molecular weight is 451.7; index of refraction, 1.735; specific gravity, 3.5 to 3.7; and hardness, 6. The color shows a wide range from white through various shades of yellow and brown to red.

Pyrope.—The formula of pyrope is $Mg_3Al_2(SiO_4)_3$. Its molecular weight is 404.6; index of refraction, 1.742; specific gravity, 3.7; hardness, 7; color, deep red to black.

Almandite.—Almandite has the formula $Fe_3Al_2(SiO_4)_3$. Its molecular weight is 499.1, index of refraction, 1.778 to 1.830; specific gravity, 3.9 to 4.2; hardness, 7 to 7.5; color, red, brown, or black. The deep red variety known as carbuncle has been used as a semiprecious stone since the earliest days of civilization.

Rhodolite.—Rhodolite is an isomorphous mixture of pyrope and almandite in the ratio of 2 molecules of pyrope to 1 of almandite; therefore its formula is $2(Mg_3Al_2(SiO_4)_3).Fe_3Al_2(SiO_4)_3$. Its index of refraction is 1.760; specific gravity, 3.80 to 3.90; hardness, 7 to 7.5; color, rose pink to dark red.

Spessartite.—The formula of spessartite is $Mn_3Al_2(SiO_4)_3$. Its molecular weight is 496.4; index of refraction, 1.800; specific gravity, 4.2; hardness, 7; color, brown to red.

Andradite.—Andradite, represented by the formula $Ca_3Fe_2(SiO_4)_3$, has a molecular weight of 509.3. Its index of refraction is 1.865 to 1.94; specific gravity, 3.85; hardness, 7; color, yellow, green, brown, or black. This variety of garnet is very common and it is an accessory mineral in many different rocks. A yellow or greenish variety of andradite is known as topazolite, an emerald green variety as demantoid, and a black variety is often termed melanite.

Uvarovite.—Uvarovite has the formula $Ca_3Cr_2(SiO_4)_3$. Its molecular weight is 501.7; index of refraction, 1.838; specific gravity, 3.5; hardness, 7 to 7.5; color, emerald green.

Schorlomite, the variety of garnet containing titanium, is too rare to be of other than scientific interest.

Specimens of garnet are usually classified under the name of the variety whose chemical composition is approached most closely. The specific gravity and index of refraction of any variety can hardly be considered constant, as both are affected by slight changes in chemical composition. The figures given in the preceding descriptions of varieties are those which have been determined for specimens whose chemical composition most closely approached the theoretical.[1]

OCCURRENCE

The members of the garnet group are common accessory minerals in a large variety of rocks. They are particularly common in the granitic rocks, gneisses, and schists, but are abundant in contact metamorphic zones and in metamorphosed crystalline limestones, as well as in such basic rocks as serpentine and peridotite. Much garnet is undoubtedly of secondary origin, having been formed by the molecular rearrangement of the mineral constituents of sedimentary rocks under heat and pressure sufficient to dissociate the original minerals. Some occurrences of garnet in granite, pegmatite, and similar rocks seem to be primary, the garnet having crystallized out of the magma solution as one of the original rock minerals. The principal minerals associated with garnet are quartz and members of the feldspar, mica, pyroxene, and amphibole groups. Most varieties of garnet are more resistant to chemical and mechanical erosion than are the associated minerals, except quartz, and for this reason garnet associated with quartz particles that have likewise undergone little alteration are common in the detritus of disintegrated rocks and in sand deposits. Garnet has a higher specific gravity than most of the minerals associated with it and consequently it tends to concentrate at the lowest part of disintegrated and sandy accumulations. This is particularly noticeable in sand and gravel deposited by streams or along the shores of lakes.

As garnet occurs in a large variety of rocks, the geographical areas in which it may be found are enormous both in extent and in number. Concentrations of garnet possessing the necessary qualifications for ornamental or industrial use, and so situated with regard to transportation and markets that they can be exploited commercially are, however, relatively small and occur in only a few areas. Because of the comparatively small demand for garnet, development of these deposits has not been large.

---

[1] See Clark, F. W., The data of geochemistry: U. S. Geol. Survey, Bull 770, 1924, pp. 404–407. Larsen, E. S., The microscopic determination of the nonopaque minerals: U. S. Geol. Survey, Bull. 679, 1921, pp. 178–180.

UTILIZATION

AS A SEMIPRECIOUS STONE

Since prehistoric times specimens of garnet possessing attractive color have been used for ornaments.

In common with other gem stones that were popular in the Middle Ages, a number of magical and medicinal properties were ascribed to this mineral. It was emblematic of constancy and was believed to possess the particular virtue of dispelling poisonous and infectious airs. It preserved health, reconciled differences between friends, and when worn suspended from the neck was said to ward off plague and thunderbolts, strengthen the heart, and increase riches and honor.

Traces of the early superstitions concerning gem stones exist to-day in the custom of wearing "birth stones" which are believed to exert a beneficial effect upon the wearer. Garnet is the birth stone for January and some people still claim that its use imparts some of its magical properties to those born in that month.

Because of their abundance and their striking color, the red and crimson varieties of garnet are most commonly used as gem stones. Garnets exhibiting other colors are more uncommon and are therefore much less utilized. Yellow and orange crystals of the hessonite variety of grossularite have furnished a few fine gems, the bulk of the production having come from gravel deposits in Ceylon. In the United States hessonite of excellent quality has been found in San Diego County, Calif.

The fiery red pyrope is undoubtedly the most popular variety of semiprecious garnet and constitutes the chief supply because of its abundance in the Bohemian Mittelgebirge near the towns of Teplitz and Aussig, which are approximately 60 kilometers north of the city of Prague in Czechoslovakia. This is probably the only place in the world in which the collection and preparation of gem garnet for the market has been of sufficient importance to approach an established industry. The diamond mines at South Africa have produced many fine pyrope garnets as a by-product. Pyrope of excellent quality is abundant in the Navajo Indian reservation in Utah and Arizona. The most important producing areas are the Mule Ear and Moses Rock fields in southern Utah, and the Garnet Ridge field in the adjacent part of Arizona. Here the garnet is scattered through the sands and gravels in small crystals that have been rounded by the scouring action of wind-blown sand. The crystals are collected from the surface by the Indians and sold in small lots to tourists and traders. It is believed that these garnets were liberated by the decomposition of a garnetiferous schist or gneiss. Pyrope has also been found in New Mexico, Kentucky, Colorado, California, Madagascar, Ceylon, Brazil, and many other widely distributed localities.

The crimson almandite has furnished some fine gems which are generally cut en cabochon because the depth of the color makes a faceted stone appear dark and lifeless. This type of garnet is more commonly known as carbuncle, and has been used extensively for ornament since very early times. Bohemia, Ceylon, Madagascar, India, Brazil, and Alaska have furnished considerable almandite of gem quality. In the United States it has been found in many localities, particularly in Pennsylvania, New York, North Carolina, Colorado, and California.

Rhodolite suitable for gems has been found in only one locality; practically the entire production has come from Macon and Jackson Counties in the western part of North Carolina. Many beautiful gems have been cut from the pale-red and rose-colored garnet obtained there. These stones are exceptionally brilliant and their pleasing color has increased their popularity.

Very little spessartite has been found in crystals having the transparency, color, and size that would make them of value as decorative stones. The only locality in the United States which has produced any appreciable amount of this stone is Amelia Courthouse, Amelia County, Va., where a few orange-brown stones of good quality have been found. Spessartite of gem quality has been reported in Nevada and a few crystals have been found in Ceylon.

Until a few years ago andradite was not considered to have any possibilities as a gem stone and it was not utilized for decorative purposes until the discovery of the green variety known as demantoid. This variety is found in Russia on the western side of the Ural Mountains, where it occurs in small crystals ranging in color from olive to emerald green. The deep green varieties were at first confused with the genuine emerald, until their true nature was revealed by chemical analysis. The brilliant luster and great color dispersion of demantoid add to its beauty as a gem.

Although uvarovite of a pleasing green color has been found, it has not been utilized as a gem stone because crystals large enough for cutting are exceedingly rare.[2] According to a recent report, considerable quantities of massive uvarovite suitable for ornamental purposes has been found in the Bushveld in Western Transvaal, chiefly on Buffelsfontein No. 205 about 6 miles from Wolhuter Station on the Pretoria-Rustenburg Railroad, 40 miles west of Pretoria. This deposit has been prospected by short trenches and shallow excavations to a maximum depth of 9 feet. Chromite is

---

[2] Hall, A. L., Chrome-garnet from the western Transvaal Bushveld; South African Min. and Eng. Jour., Sept. 27, 1924, pp. 63–64.

intimately associated with the uvarovite. The occurrence is of enough promise to warrant exploitation and the product is now being marketed under the trade name of South African "jade." It is homogeneous and even textured and no individual minerals can be identified with the naked eye except where the chromite is present. The range of color is wide; shades of green are most common, and some pink, white, and bluish specimens have been found. The hardness, which ranges from 7 to 8, enables the "jade" to take an excellent polish when worked. The deep-green translucent varieties are most valuable because of their resemblance to the true jade, which is highly prized by the Chinese. Several lots have been sold in the Far East at a figure in excess of that paid for crude Burmese jade.

It has been estimated that in 1922 the world's production of gem garnet was worth $68,000.[3]

MINING OF GEM GARNET

Mining for gem garnet has never developed into a systematic, well-established industry in any one locality for two reasons: (1) The enormous geographic distribution of garnet deposits tends to localize the market and to prevent any one deposit from supplying the bulk of the world's demand, and (2) the present buying capacity of the public can be satisfied by a few small and scattered operations. Garnet is one of the cheapest gem stones and commands so low a price that its extraction is seldom profitable. The small amount of gem garnet produced annually is obtained from small quarries that are worked irregularly or is collected from superficial deposits formed by the disintegration of garnetiferous rocks. In the quarries the rock is drilled by hand and shot with small charges of explosives, so that it can be removed and be broken down carefully by hand to liberate the garnet crystals without shattering them. As its shattering effect is less, black powder is preferable to dynamite as an explosive. Quantities of garnet have been freed by the weathering and disintegration of garnetiferous rocks. Considerable garnet of gem quality has been collected from loose sand and from stream gravels in which this liberated garnet had concentrated. A few crystals of garnet of gem quality are occasionally recovered as a by-product in the mining of abrasive garnet and in the operation of gold placers.

PRODUCTION IN UNITED STATES

Some garnet was undoubtedly used for ornament by the early European settlers in America. The literature of colonial days contains very few references to the mineral, and interest aroused in it probably did not extend beyond the occasional collection of a crystal

[3] Ball, Sydney H., The geologic and geographical occurrence of precious stones: Econ. Geol., vol. 17, November, 1922, pp. 575–601.

of unusual perfection or attractive color. One of the early references to American garnet is contained in "An Essay about the Origine and Virtues of Gems," by the English chemist, Robert Boyle, which was published in 1672. In discussing the origin of minerals Boyle says:

> I have some American granats which I had a great and peculiar reason to believe had been once liquid bodies, and therefore thought them the more worthy to be examined; and finding their colour to be so deep that they were almost opacous, and by judging by my hand that they were much heavier than pieces of cristal (quartz) of the same bulk would be, I weighed them in a pair of nice scales in the air and in the water, and found them, as I had expected, to be almost four times as heavy as water of the same bulk.

This specific gravity determination indicates that the garnet in question was probably almandite.

The records of the production of gem garnet in the United States indicate the unimportance of this market, as the annual output has seldom had a value exceeding a few thousand dollars.

Value of gem garnet produced in the United States

[Statistics from the U. S. Geological Survey]

1883	$6,000	1896	$2,600	1909	$1,650
1884	4,000	1897	9,000	1910	3,100
1885	2,700	1898	7,000	1911	2,065
1886	3,250	1899	7,000	1912	860
1887	3,500	1900	21,500	1913	4,285
1888	3,500	1901	22,100	1914	1,760
1889	2,308	1902	2,500	1915	4,523
1890	2,308	1903	3,000	1916	1,542
1891	3,000	1904	3,000	1917	624
1892	5,250	1905	5,000	1918	1,277
1893	2,000	1906	3,000	1919	1,630
1894	4,300	1907	6,460	1920	331
1895	2,350	1908	13,100	1921	606

In 1900 rhodolite valued at $20,000 was produced in North Carolina; in 1901 the production was valued at $21,000; this rhodolite accounts for the unusually high figures for the two years. Since 1901 the production of garnet has declined to a figure so unimportant that no records have been kept since 1921.

AS A JEWEL FOR BEARINGS

Garnet has been used to a limited extent as a jewel in the bearings of watches, meters, and scientific apparatus. The material employed for this purpose must be hard enough to withstand the wear caused by the friction of moving parts and tough enough to be manufactured into thin slices that will not shatter under rough handling. Most of the garnet used for bearings consists of the chips produced during the cutting of gem garnet from Bohemia and Madagascar. These chips are ground down until very thin, given a rounded form, and then pierced in the center with a small drill. Before the jewels are set in a watch the opening is redrilled to make it fit accurately the axle which rotates on this bearing. Sapphire and ruby—the blue and red varieties of corundum—are considered better than garnet for bearings, as they are harder and consequently more durable, and they are therefore used more extensively. Millions of these jewels are annually used in watch movements. Garnet is used only in the cheaper movements, but the number of garnet jewels used monthly in watches is estimated at 250,000. Jewels are made in Switzerland, which supplies the greater part of the world's requirements. Aside from a very small production during the World War, when imports were curtailed, no garnet for jewel purposes has been produced in the United States. The supply from foreign sources has been abundant and has sold at a price below the cost of domestic production.

FOR THE MANUFACTURE OF FERROSILICON

The elements that compose garnet are not of enough value to make their extraction profitable. Extraction would necessarily be difficult and expensive because of the stability of the silicates composing the garnet group. As far as is known only one attempt has been made to utilize the chemical composition of garnet. United States Patent 1192394, issued July 25, 1916, describes a process for the manufacture of ferrosilicon and a highly aluminous abrasive from almandite garnet. The patentee proposed to utilize the considerable quantity of garnet flour produced in the manufacture of garnet abrasives. This powder was too fine for use as an abrasive, and there was no market for it in 1916. The process consisted in the reduction of the garnet in an electric furnace. The furnace charge was garnet crushed to one-sixteenth inch or less, mixed with approximately 18 per cent of its weight of carbon. An alternating current of low voltage and high amperage was supplied to the electrodes, one set in the base of the furnace, the other, a graphite rod that could be raised or lowered, suspended vertically in the center of the charge. The amount of carbon added to the charge was sufficient to reduce all the constituents of the garnet except the alumina; enough silica was added to insure the conversion of the iron content of the garnet to a ferrosilicon alloy. The charge in the furnace was heated to the fusion point of alumina and maintained at this temperature until all of the garnet constituents except alumina were reduced. Then the charge was cooled slowly, so that the alumina could crystallize and the ferrosilicon solidify in nodules.

A description[4] of the experimental work employing this process says that a ferrosilicon alloy containing 24 to 25 per cent silicon and an aluminum abrasive containing about 62 per cent alumina could be produced from the original garnet.

The process was devised when ferrosilicon was expensive and difficult to obtain because of war-time conditions. It is doubtful if the process would be feasible at present, as ferrosilicon is abundant, whereas garnet is too expensive and is hardly sufficient to maintain such a process on a commercial scale.

AS AN ABRASIVE

Present activity in the mining of garnet has resulted from recognition of the unusual abrasive properties of the mineral. The bulk of the garnet mined for use as an abrasive has been almandite; rhodolite, pyrope, and andradite have been utilized to a lesser extent. Deposits of the other varieties which could produce a commercial tonnage of abrasive garnet are unknown. Abrasive garnet is utilized either in the form of a manufactured paper, similar to sandpaper, or as a loose grain or powder for grinding and polishing. The average hardness of abrasive garnet is 7.5, but some specimens have shown a hardness approaching 8.0. Quartz that is used for the manufacture of ordinary sandpaper is not as hard; it has a hardness of 7.0. The marked lack of any regular cleavage and the manner in which the mineral fractures affect the shape of the grains of crushed garnet and thereby influence its industrial use.

Although much garnet has a laminated structure, which furnishes parting planes along which the mineral separates when crushed, this structure can not be considered true cleavage because the garnet shows no further tendency to part regularly when it has once been separated into the plates formed by these laminations. Furthermore, this parting is purely mechanical and is not related to the crystallization. Garnet used as an abrasive has a fracture that is irregular to subconchoidal, consequently the grains of crushed garnet are irregular, many-angled particles which are roughly equidimensional and suggest modified cubes or tetrahedrons which have a multitude of sharp chisel-like cutting edges. As these grains break in use further sharp edges are produced. Garnet seems to have just enough brittleness to break under the strain of ordinary use rather than to wear down to a smooth surface.

Substances having a distinct conchoidal or shell-like fracture, of which glass is a typical example, shatter into fragments that tend to be flat and very thin. Although these flat particles retain very

[4] Thompson, M. de Kay, and Davenport, John, The electric furnace reduction of garnet: Chem. and Met. Eng., vol. 22, 1920, p. 596.

sharp cutting edges, it is impossible to arrange them on paper in such a manner that they will make an efficient abrasive. Either they tend to lie flat on the paper or else they project in long splinters which make deep scratches in the material being ground or polished. These scratches are highly objectionable, as their removal requires further grinding and polishing and thus increases the cost. Particles of garnet have enough flat surfaces to permit firm attachment to the paper backing; at the same time they expose a number of cutting edges which are practically in the same plane and therefore abrade evenly. Powdered garnet is used chiefly for grinding plate glass; here its greatest asset is its hardness, which enables it to abrade the softer glass with rapidity.

DEVELOPMENT OF THE ABRASIVE GARNET INDUSTRY

The initial developments in the mining of garnet were due to the perception of the late Mr. H. H. Barton, of Philadelphia, Pa. According to a letter from Mr. C. R. Barton to the writers, he first recognized the abrasive properties of the mineral and realized their commercial importance. It is interesting to note in this connection that the utilization of garnet as an abrasive is in a degree an outgrowth of its use as a gem. During the period from 1860 to 1865, while Mr. Barton was in business as a jeweler in Boston, Mass., his attention was attracted to garnet by a gentleman who brought in a small amount to be appraised for its gem value. This garnet was from the Adirondack section of New York State. Although the mineral was of good quality, it was thought that the market was too small to permit a domestic product to compete with the importations of gem garnet from Bohemia, so the matter was dropped and for the time was forgotten.

Some years later Mr. Barton engaged in the manufacture of sandpaper from crushed quartz, flint, and, to a lesser extent, glass. Keen competition resulted in the investigation of possible improvements. For a time a new abrasive paper known as "ruby paper" was manufactured, in which the quartz was replaced by red carnelian, a form of silica closely related to agate. This carnelian was collected in California. The supplies were limited, the delivery uncertain, and eventually the use of carnelian as an abrasive was abandoned.

In continuing the search for an improved abrasive the possibilities of garnet were investigated. Its superior qualities were quickly recognized and a search was begun for a deposit capable of commercial production. Mr. Barton remembered the gem garnet from the Adirondacks of New York, and located the deposit from which the small sample had been taken years before. As this deposit proved to be very large and rich in garnet, the manufacture of garnet paper on a commercial scale was undertaken. Garnet paper was first made about 1880 and began to be of commercial importance about 1882.

Contemporaneous with this development Herman Behr & Co. (Inc.) opened a deposit of garnet near Boothwyn, Delaware County, Pa. This garnet occurred in a mica schist which extended from the surface to a depth of 30 or 40 feet. The garnet content was high, averaging 30 per cent of the rock mass. The surface of the rock was so disintegrated that down to 20 feet the garnet could be extracted readily. Below 20 feet the rock was unaltered, the separation of the garnet became more difficult, and milling operations were started. This was the first attempt to mill garnet ores, as the previous output had been obtained by hand cobbing the disintegrated rock. The harder garnet-bearing rock mined later was broken by mulling and placed in a washing machine that separated the garnet from the dirt and gangue. Operations at Boothwyn were abandoned in 1898 as garnet of better grade became available from other sources. A small quantity of garnet was also produced by underground mining in a decomposed gneiss near Chester Heights, Delaware County, and by small workings in Chester County, Pa.

The recognition of the value of abrasive garnet resulted in the search for other deposits and operations were started in Connecticut, North Carolina, and New Hampshire. The Connecticut deposits, near Roxbury and Roxbury Falls, were eventually abandoned after being worked for a time on a small scale and New York, New Hampshire, and North Carolina remained the only producing States.

ADVANCES IN MINING AND MILLING

The first milling of garnet ores on a comparatively large scale was done by Mr. F. C. Hooper, who established the North River Garnet Co. at North River, N. Y. Separation of garnet from associated minerals by mechanical means on a large scale had received little attention until 1893, when the first mill of the North River Co. was erected. The principles and equipment of standard metal practice were adapted to meet the particular requirements of garnet milling, and special machines were designed to treat such special mill products as could not be properly handled with the equipment available. Accompanying the development of better equipment for milling garnet ores came more systematic methods of mining. Large quarries were opened, hand drilling and loading were abandoned, and mechanical equipment replaced the older and slower methods of handling the ore. New York now holds a dominant position in the production of garnet because of the early development of the industry in that State and the abundance of high-grade ores in the Adirondack district. Garnet-bearing rocks occur over a wide area in Warren, Essex, and Hamilton Counties and especially in the district adjacent to the intersection of the boundaries of these counties. The garnet is in fairly well-developed crystals and also in large massive aggregates showing little trace of individual crystallization. Most of the rocks are gneisses and have undergone extensive metamorphism. Their origin is uncertain. In some localities the rocks seem to be much altered sediments, in other places they seem to be of igneous origin. The garnet content of the rocks seldom exceeds a maximum of 12 per cent, although the massive aggregates are occasionally found in large masses containing as much as 60 per cent garnet. In spite of the wide distribution of the rocks, garnet concentrations rich enough to make the rock commercially important are not common and are found in comparatively few localities.

The present production of this district is limited to the operations of four companies—the North River Garnet Co., Barton Mines (Inc.), the Warren County Garnet Co., and the American Glue Co. In the past a number of small scattered workings have produced irregularly but none of them are active to-day.

DESCRIPTIONS OF MINES AND MILLS

NEW YORK

THE NORTH RIVER GARNET CO.

The operations of the North River Garnet Co. are on the eastern side of Thirteenth Lake, in Warren County, N. Y., about 10 miles from the village of North Creek, which is the terminus of the Adirondack branch of the Delaware & Hudson Railroad and the nearest shipping point. Motor trucks running over State and privately owned roads carry in supplies and bring the garnet concentrates from the mill to the railroad for shipment.

QUARRY

The garnet, which approaches almandite in composition, occurs in a gneiss which apparently is a metamorphosed sediment of the Grenville series. The garnet content of this gneiss ranges from 4 to 8 per cent, and is in crystals that attain a maximum diameter of 3 inches, but average about five-eighths inch. Hornblende and feldspar are the most important gangue minerals, but pyroxene, mica, magnetite, pyrite, and ilmenites are present in small amounts. A large projecting knob of this gneiss has been opened by a pit 300 feet in maximum diameter. The face of this quarry is 140 feet high at the highest point. The rock shows highly developed joint systems (see Pl. 1, *A*) and the horizontal partings are utilized for the base of small benches in quarrying. Systematic large benches are not developed. Shot holes are drilled with power drills to a depth of 15 feet. The holes are sprung and then loaded with 60 per cent ammonia dynamite, which is fired with a blasting machine. The jointing of the rock tends to produce large blocks and considerable blockholing is necessary. The drilling for this is done with machine drills and the holes are loaded with 40 per cent dynamite. Car tracks, 36-inch gauge, run from the working faces to the head of the mill, a maximum distance of 200 yards, on a grade that permits the loaded cars to run by gravity. A horse draws back the empties; the broken rock is either loaded by hand into a double open-end car with a capacity of 8 tons or is moved with a steam shovel. Two power shovels with dippers having a capacity of seven-eighths of a yard are used; they load the rock into special cars with a capacity of 8 tons. Severe weather interferes with the operation of the quarry in winter.

MILL

The mill,[5] which stands on a hillside just below the quarry, has a capacity of 500 tons a day. Milling consists essentially in liberating the garnet from the gangue minerals by stage crushing and concentrating it with jigs. Harz and James jigs are used for the coarser sizes; the fines are recovered on vanning and pneumatic jigs specially designed for this purpose by Mr. F. C. Hooper. Figure 1 shows the flow sheet of the mill. Dry, stage crushing with jaw crushers and rolls reduces the rock to fragments one-quarter to five-sixteenths inch in size which go to a storage bin with a capacity of 200 tons. From this bin the ore is fed by water to Harz jigs pulsating two hundred times per minute and provided with one-quarter and three-sixteenth inch screens. These jigs make a clean tailing, which immediately goes to waste, and a middling hutch product which, after being partly dewatered in cones, goes to four James jigs pulsating two hundred and twenty-five times per minute, and one Hooper vanning jig. These jigs, all provided with 6-mesh screens, make a clean tailing, a middling, and hutch product. The hutch product goes to a series of vanning jigs so arranged that the hutch product from one jig is the feed for the next. The screens of these jigs are 12, 16, 22, and 30 mesh. Except the jig having a 12-mesh screen, the jigs produce a clean concentrate which is skimmed

---

[5] For a complete description of this mill see Wormser, F. E., Mining, concentrating, and marketing garnet: Eng. and Min. Jour.-Press, vol. 118, Oct. 4, 1924, pp. 525–531.

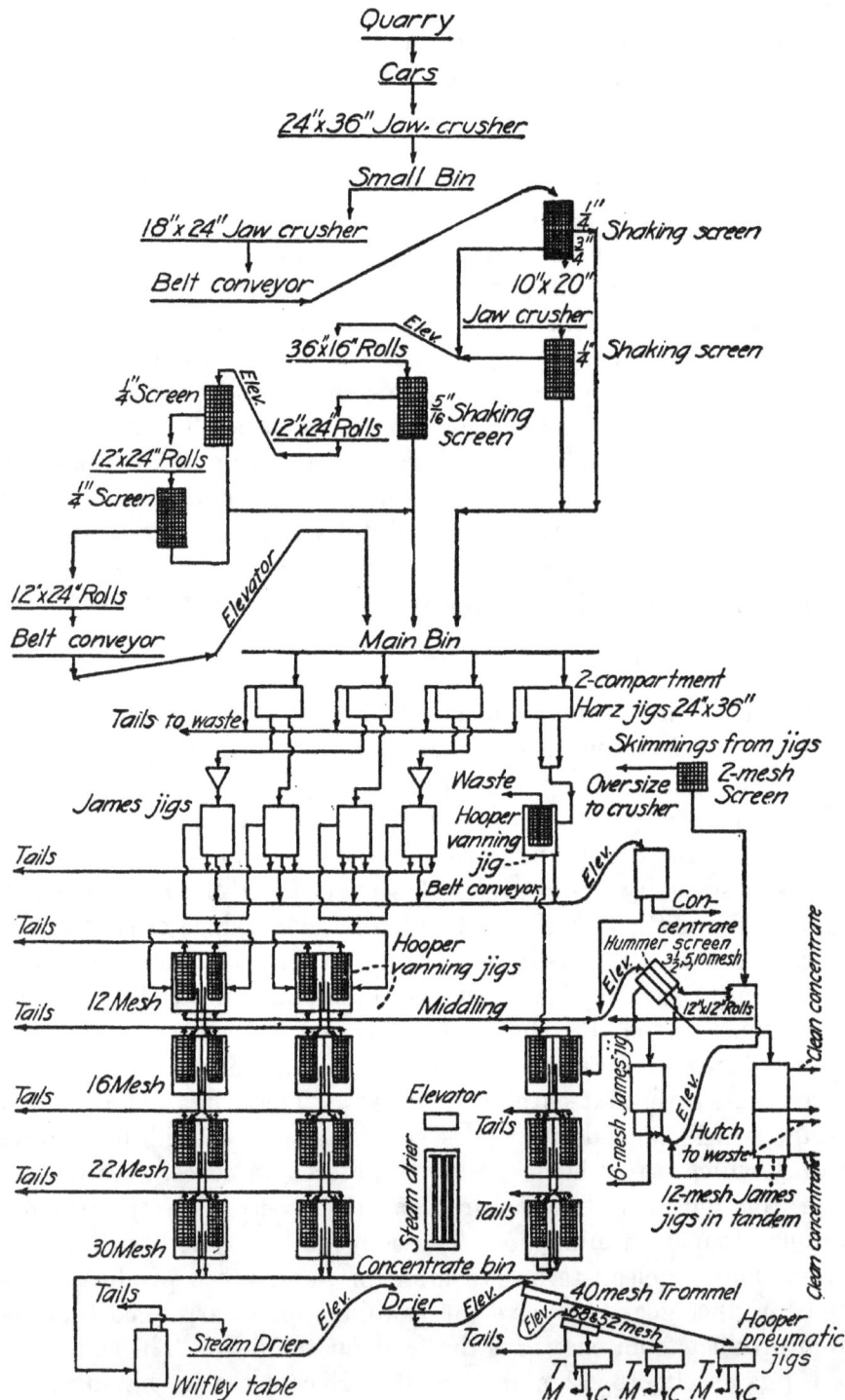

FIGURE 1.—Flow sheet of mill of the North River Garnet Co.

and removed by hand. The middling produced is sent to a Wilfley table which makes a clean tailing, which is sent to waste, and an enriched middling. This middling goes by an elevator to a steam drier; after drying it is screened through 40, 52, and 68 mesh screens. The minus 68-mesh material goes directly to waste. The other products are concentrated on pneumatic jigs which produce a concentrate, middling, and tailing. The middling is returned to the pneumatic jigs.

An elevator takes the middling from the four James jigs and the single Hooper vanning jig to a single-cell James jig which makes a clean concentrate and a middling which goes to a set of Hummer screens. These screens size the material to 3½, 5, and 10 mesh. The middling product from the vanning jigs having a 12-mesh screen also goes to these screens. Rolls crush the oversize from the 3½-mesh screen and also the skimmings from the Harz jigs which have passed a 2-mesh screen. These rolls are in closed circuit with the Hummer screen. The plus 5 mesh goes to a James jig having a 6-mesh screen, which makes a concentrate and middling. The plus 10 mesh goes to a two-cell James jig, provided with 12-mesh screens, which makes a concentrate, middling, and tailing. The middling from the 6-mesh and 12-mesh jigs is recrushed in the rolls and sent to the Hummer screen. The minus 16 mesh goes to the 16-mesh vanner jig where it joins the circuit of the 22 and 30 mesh jigs.

An elevator carries all the concentrates to an inclined steam drier which discharges into a bin that receives the mixture of all the different concentrate products. This final concentrate averages 90 per cent garnet. It is sacked and then trucked to North Creek for shipment by rail. Three 150-horsepower coal-fired Stirling boilers supply steam to a tandem Corliss-valve engine which drives the mill.

Part of the tailings are utilized for road construction. They also make an excellent concrete aggregate but are not valuable enough for this purpose to warrant transportation beyond the immediate locality.

THE BARTON MINES CORPORATION

The mining and milling operations of the Barton Mines Corporation are situated on Gore Mountain at an altitude of 2,800 feet, about 4½ miles southwest of the village of North Creek, N. Y. By road the distance is approximately 11 miles and in places the grade is high, as there is a rise of 1,800 feet in approximately 3 miles. Trucks bring in supplies for the camp and carry the garnet concentrates to the railroad.

QUARRY

The ore body has a general east-west strike, and its outcroppings have been traced a total length of more than 4,000 feet; its total extent is unknown. The overburden is slight. Quarrying operations have exposed the ore body for a length of 2,000 feet and a breadth of 150 feet.

The garnets occur in a metamorphosed igneous rock of uncertain origin but seemingly allied to syenites that are common in the vicinity. The rock shows little banding or schistosity, its most marked feature being the garnet crystals which give it a porphyritic texture. The mineralization is simple. Hornblende constitutes nearly 40 per cent of the mass; almost all the remainder is divided between orthoclase and plagioclase feldspars, pyroxene and biotite. Small amounts of magnetite, pyrite, and ilmenite are present. The garnet content of the ore body averages 12 per cent in the areas exposed by quarrying. The garnet occurs as imperfectly developed crystals, locally called "pockets," which attain remarkable size. Single crystals 1 foot in diameter are common; crystals 30 to 36 inches in diameter have been found and have yielded over a ton of garnet. Most of the crystals are surrounded with thin shells of pure hornblende, to which are attached masses of pale green plagioclase feldspar, approaching oligoclase in composition, which indicate fractional crystallization of the rock as it solidified. Conditions must have been exceptionally favorable to prolonged crystal growth to permit the development of crystals of a size unparalleled in other deposits. These garnet crystals have a decided laminated structure by which they are readily divided into plates from one-sixteenth to one-quarter inch thick. A thin film of pyrite or other mineral often lies in the lamination. The garnet approaches almandite in composition.

Weathering has altered the garnetiferous rock near the surface and the oxidized zone has developed to a depth of 15 feet. In this oxidized zone the feldspars are completely kaolininzed, and the hornblende has been altered until a siliceous, iron-stained residue is left. The garnet in the oxidized zone shows practically no indications of change other than a tendency to break readily along the lamination planes, which have been weakened by the alteration of the separating films of pyrite and other minerals.

The first mining, which began over 40 years ago, was in this oxidized zone. Because the ore was crumbly it could be readily worked with pick and shovel and the garnet extracted by hand. Mining operations were carried on irregularly by leasers who sold their productions to buyers of abrasives. The production was small, not more than 900 tons a year. At a later date mining of the unoxidized rock began, and when the erection of a modern concentrating mill was completed in March, 1924, this mining was systematized.

The quarry now consists of a series of open pits which are being developed into regular benches. Mining is most active on two faces, one of which attains a maximum height of 22 feet. Ten-foot holes, drilled with jack hammers, are blasted with 40 and 60 per cent gelatine. More drilling is done on the bench developed by shooting the first set of vertical holes, and when these are shot the face is advanced. A large number of drill holes, totalling several hundred feet, are drilled in the floor of the quarry back of the advance face. Bowlders are block-holed with small charges of dynamite. In breaking the ore many of the garnet crystals are shattered. The largest fragments of clean garnet, which amount to several hundred pounds a day, are picked up and sacked immediately for shipment.

The broken ore is loaded into 3-ton side-dump cars with a steam shovel mounted on caterpillar treads and having a capacity of three-fourths cubic yard. A gasoline locomotive hauls four-car trips to the mill 200 yards distant. Plate I, *B*, gives a view of the mill. Plate II shows the flow sheet of the mill.

MILL

At the top of the mill the cars dump into a chute with a slope that is less than 45 degrees to provide a dirt floor and avoid abrasion. From this chute the ore passes to a 24-inch by 36-inch Blake-type jaw crusher with a 3-inch gap. Railroad rails suspended just in front of the crusher check the rush of the largest bowlders. A short belt feeder takes the crusher discharge and drops it into an elevator which delivers it to a double trommel, the inner jacket of which has 1¾-inch holes and the outer jacket ¾-inch holes (round). The undersize passes to the storage bin; the oversize goes to a picking belt where the large pieces of garnet are removed and also much tramp material. A magnetic pulley at the discharge end of this belt removes tramp iron. The belt discharges into a gyratory crusher which is in closed circuit with the elevator and the double trommel. Hence, the storage bin contains ore that is all through three-fourths inch. Although the grinding is presumably dry, the moisture content of the ore is such that there is very little dust. As the capacity of the bin is about 500 tons it is proposed to do all of the crushing on the day shift and run only the concentrating part of the mill 24 hours a day.

The storage bin discharges through five hoppers to a flat conveyor, driven by a ratchet arrangement, which delivers the ore to the main feed elevator. At the end of this belt water is added for the first time. The elevator discharges to a chain drag which dewaters the material before it passes to a series of five trommels. The overflow of the drag goes through a chip catcher to the Dorr thickener. The five trommels are arranged in this order: One-half inch, three-eighths inch, one-fourth inch, one-eighth inch, and $2\frac{1}{2}$ millimeters. The chain drag discharges first to the one-half-inch trommel, and the oversize from that goes to a 30-inch by 16-inch set of rolls which is in closed circuit; consequently, all of the ore passes through one-half-inch round hole before concentration begins. The undersize from the one-half-inch trommel passes to the three-eighths-inch trommel, and the undersize of that to the one-fourth inch. The four sizes of concentrates made by the series of trommels are each fed to three-compartment jigs. The first size, minus one-half inch plus three-eighths inch, is fed to two of these three-compartment jigs, as is also the second size, minus three-eighths inch plus one-fourth inch. The third size, minus one-fourth inch plus one-eighth inch, is of such volume, however, that three of the three-compartment jigs are required to treat it. The finest size, minus one-eighth inch plus $2\frac{1}{2}$ millimeters, is so small in quantity that it requires only one three-compartment jig. Eight three-compartment jigs are required to treat these four sizes. The jigs run at high speed with a short stroke. The depth of bed ranges from about $1\frac{3}{4}$ inches to about $3\frac{1}{2}$ inches. The jig sieves are light trommel sheets with round-hole openings about $1\frac{1}{2}$ millimeters to 3 millimeters in diameter. The jigging area in each compartment is about 24 inches by 36 inches. The first two cells of each of these eight jigs produce clean or finished concentrates over the cup draw; the third cell produces a finished tailing, which goes to the tailing elevator, and an unfinished middling over the cup draw. All of these middlings pass back to the main feed elevator. The middlings from the jigs treating the coarser sizes pass to this elevator through the rolls grinding the oversize from the one-half-inch trommel.

The undersize from the $2\frac{1}{2}$-millimeter trommel passes to a cone where it is dewatered; the overflow passes to the Dorr thickener, through the chip catcher, and the sands are fed by a distributor to four tables, which produce a finished tailing, a middling that goes to the elevator feeding the second set of tables, and a low-grade dirty concentrate which is fed without dewatering to four small one-compartment jigs; these jigs produce a finished concentrate through the cup draw, an unfinished tailing over the sides of the cell, and an unfinished hutch product through the jig sieve or bed. Part of the bedding on these jigs is small bird-shot. The tailings and the hutch product are combined and join the table middlings from the first set of tables; they are elevated, dewatered in a cone, and fed by a distributor to a second set of four tables which produce a finished tailing, an unfinished or dirty concentrate which is sent to the four small jigs for final cleaning, and a middling which is returned to the elevator handling this table feed. Thus these two sets of tables and the four small jigs are in a double circuit. The tailings from the eight tables and from the eight jigs treating sized feeds are elevated and dewatered in a duplex classifier, the discharge of which passes to the tailing pile, and the overflow goes to the Dorr thickener, measuring 40 feet by 10 feet through the chip catcher, an ingenious arrangement consisting of a trommel sheet and scrubbing brushes on a chain drag for removing wood pulp, waste, etc. The spigot product of the Dorr thickener goes to waste, but all of the overflow water is reused in the mill. The concentrates from all the jigs are elevated and dewatered in a simplex classifier. The dewatered concentrates pass to an oil-fired drier. The dried product is weighed into 100-pound sacks and stored in the basement until shipped. Trucks haul the finished concentrates to North Creek for rail shipment to manufacturers of abrasives. A ball mill is being installed to mill the finer sizes, for which there is a small market, to abrasive powder suitable for grinding plate glass. The tailings are used to repair the roads. They would make a satisfactory concrete aggregate but no local market is available.

Two 130-horsepower vertical two-cylinder oil engines furnish power. One drives the air compressor and the crushing or dry part of the plant; the other drives the wet or concentrating part of the mill and a 60-kilowatt generator. Either engine can be used to drive either part of the mill—a very flexible arrangement. The oil fuel flows to the engines from a tank outside the building.

In the summer months the mill is hard pressed for water. Virtually all of the water used is reclaimed by means of the Dorr thickener. The tailings are almost dry when discharged just outside the mill. Fresh water is used only for the engines and the crusher jackets and compressor. During the winter, or for about eight months of the year, the water requirements can be met with fresh water from the mountainside.

WARREN COUNTY GARNET MILLS (INC.)

The Warren County Garnet Mills (Inc.) works a number of small, scattered quarries near Wevertown and Johnsburg, N. Y. The present output is obtained mainly from the Armstrong farms in Johnsburg. The garnet occurs in broad bands in a biotite gneiss. Much of it is in massive aggregates, which in places constitute nearly the whole rock mass. Feldspar, biotite mica, pyroxene, and hornblende are associated minerals. In some other deposits that have been worked irregularly the garnet is in well-developed crystals.

Mining is by excavations that seldom exceed 8 feet in depth. The rock is drilled by hand with a single jack, blasted, sledged, and hand picked. The rich garnet ore is separated and piled in walls so that the tonnage can be determined by measurement and is hauled to the mill by trucks. The garnet content of the ore going to the mill is high, ranging from 30 to 60 per cent.

The ore is concentrated in a dry mill with a capacity of 3 to 5 tons of concentrate a day according to the garnet content of the ore. At the mill the ore is fed by hand to a small jaw crusher from which it goes to two sets of rolls in series where grinding through about 16 mesh is completed. Screens for removing the fines are in place in front of each set of rolls. From the rolls the ore passes to a Keedy sizer which screens it into 16 sizes between 16 and 200 mesh. The 16-mesh oversize is discarded as it is mainly flaky biotite which can not be ground readily. Each of the sizes is treated separately on pneumatic tables, which are adjusted as to speed, length of stroke, and air pressure for each size handled. The concentrates and the middlings from the tables are marketed; the tailings go to waste. The middlings are sold to produce low-grade abrasives. A home-made sizer is often used to "true" or "prove" the products made by the Keedy sizer. The biotite tailing is kept separate but the quantity is so small that it does not constitute a source of supply for roofing material. The finished concentrate is shipped in kegs to wood-working industries and to manufacturers of abrasives.

THE AMERICAN GLUE CO.

A band of garnetiferous gneiss outcropping on Casey Mountain about 5 miles northwest of North River, N. Y., the nearest post office, has been operated by the American Glue Co., and has produced a considerable tonnage of garnet. In August, 1924, the mill was not active and only a small tonnage was being hand cobbed and sacked for shipment at the quarry.

The garnet (almandite) occurs in the gneiss as small crystals from $\frac{1}{2}$ inch to 3 inches in diameter, which form 4 to 8 per cent of the rock. The outcrop of the gneiss has been traced about 2,000 feet; it is partly hidden by a shallow overburden of loose soil. In addition to an open quarry there is an underground mine. A tunnel over 800 feet long was driven from near the mill to a point below the quarry floor with which it was connected by raises. The raises were filled with broken ore from the quarry and drawn when bad weather made quarrying impossible. A considerable tonnage was also obtained from underground stopes. The rock stood well and no timbering was necessary but the cost of underground mining was considerably higher than that of quarrying.

At the mill, which follows the general plan of other garnet concentration mills in this district, the ore was crushed to minus five-eighths inch plus one-half inch and treated in Harz jigs. These jigs gave a finished tailing, which was sent to waste, and a concentrate, middling, and hutch product. The middling and hutch product were crushed in rolls and screened to sizes ranging from minus ½ inch to minus 3 millimeters. These sized products were treated in a series of jigs and Wilfley tables. The finished concentrate was dried in a steam drier, sacked, and hauled in trucks to North Creek, 11 miles distant, for railroad shipment.

OTHER MINES

A number of garnet deposits in the Adirondacks have been worked irregularly in the past but are inactive to-day. They are scattered over a wide area in Warren, Essex, and St. Lawrence Counties. In Essex County deposits near Keeseville and Minerva have yielded a small output. Small workings near North Creek, North River, Riparias, and Warrensburg, Warren County, have also produced a limited tonnage. One mine near Gouverneur, St. Lawrence County, was active for a time. At these deposits the garnet was taken from irregular excavations and no concentrating mills were erected.

The active producers of abrasive garnet in the Adirondacks control, by lease or ownership, large reserves of garnetiferous rock in addition to the deposits from which they get their present output.

NEW HAMPSHIRE

WAUSAU ABRASIVE CO.

The garnetiferous rock worked by the Wausau Abrasive Co. outcrops near the top of Currier Hill in North Wilmot, Merrimac County. The garnet (almandite) occurs in small crystals one-fourth to three-eighths inch in diameter. These small crystals constitute 40 to 60 per cent of the rock. Feldspar and biotite are the most prominent gangue minerals.

There is practically no overburden and quarrying is done in the exposed outcrop. The quarry workings are irregular because inclusions of barren rock must be avoided. The present quarry measures approximately 100 by 180 feet with a maximum depth of 25 feet. Holes are drilled to a depth of 6 feet with jack hammers and shot with 60 and 75 per cent gelatin dynamite. The bowlders are bulldozed and loaded by hand on cars with a capacity of 2,400 pounds. A hoist pulls the car up an incline from the quarry floor, and then they are trammed by hand about 300 feet to an ore bin on the edge of the hillside above the mill.   From the bin the ore is fed to a 6-inch by 10-inch jaw crusher with a 1¼-inch gap.   The crushed rock falls into a loading bin from which an aerial tramway, 1,200 feet long, conveys it to the mill bin 225 feet below.

All concentrating operations are dry and consist essentially of stage crushing, screening, and concentrating the sized products on tables.   Figure 2 gives the flow sheet of the mill.   The ore from the mill bin is crushed in 26-inch by 15-inch rolls to three-eighths inch, and then elevated to a direct-heat drier.   If the ore is already dry it can be sent directly to the screens by means of a screw conveyor, and in this way the capacity of the mill can exceed the tonnage the drier can handle in 24 hours.

The discharge from the drier goes to a Newago 4-mesh screen. The undersize, under 2 millimeters, is sent to a bin and the oversize is recrushed in 26-inch by 15-inch rolls, set to crush to 2 millimeters. The product from these rolls joins the minus 2-millimeter product from the 4-mesh screen.   This 2-millimeter product is elevated to another Newago screen, divided into two sections having 6-mesh and 10-mesh screens.   Two oversize products are made, one minus 2 millimeters, plus 1.4 millimeters; the other minus 1.4 millimeters, plus 0.75 millimeter.   The undersize, minus 0.75 millimeter, goes to another 24-mesh Newago screen.   The oversize, minus 0.75 millimeter plus 0.36 millimeter, is separated and the minus 24 mesh is passed to a 48-mesh screen which separates a minus 24-mesh plus 48-mesh product that is saved for concentration.   The minus 48-mesh material goes directly to waste.   The sized products are stored in bins and are treated alternately on two dry tables that are adjusted to concentrate the different size products.   Tailings go directly to waste; middlings return to the 6-mesh and 10-mesh screens.   The concentrates go to the sacking room, in the basement of the mill, where they are put in 100-pound canvas bags; then they are hauled by truck 3 miles to South Danbury for railroad shipment.   The present output of the mill, when running one 10-hour shift a day, is 50 tons of concentrate a week.   Severe weather in winter interferes with operations and production is curtailed between December and April.

A small Hardinge mill in closed circuit with a 30-inch air classifier has been installed to make garnet powder, 200 mesh to 325 mesh or finer, for use in the fine grinding of plate glass and similar substances.   Any excess of any size in small demand is utilized in this way.   In August, 1924, a grading plant was being installed at the mill to prepare garnet that could be shipped instead of the concentrate of mixed sizes.

In addition to the active quarry operation, the company owns other deposits which provide a considerable reserve.

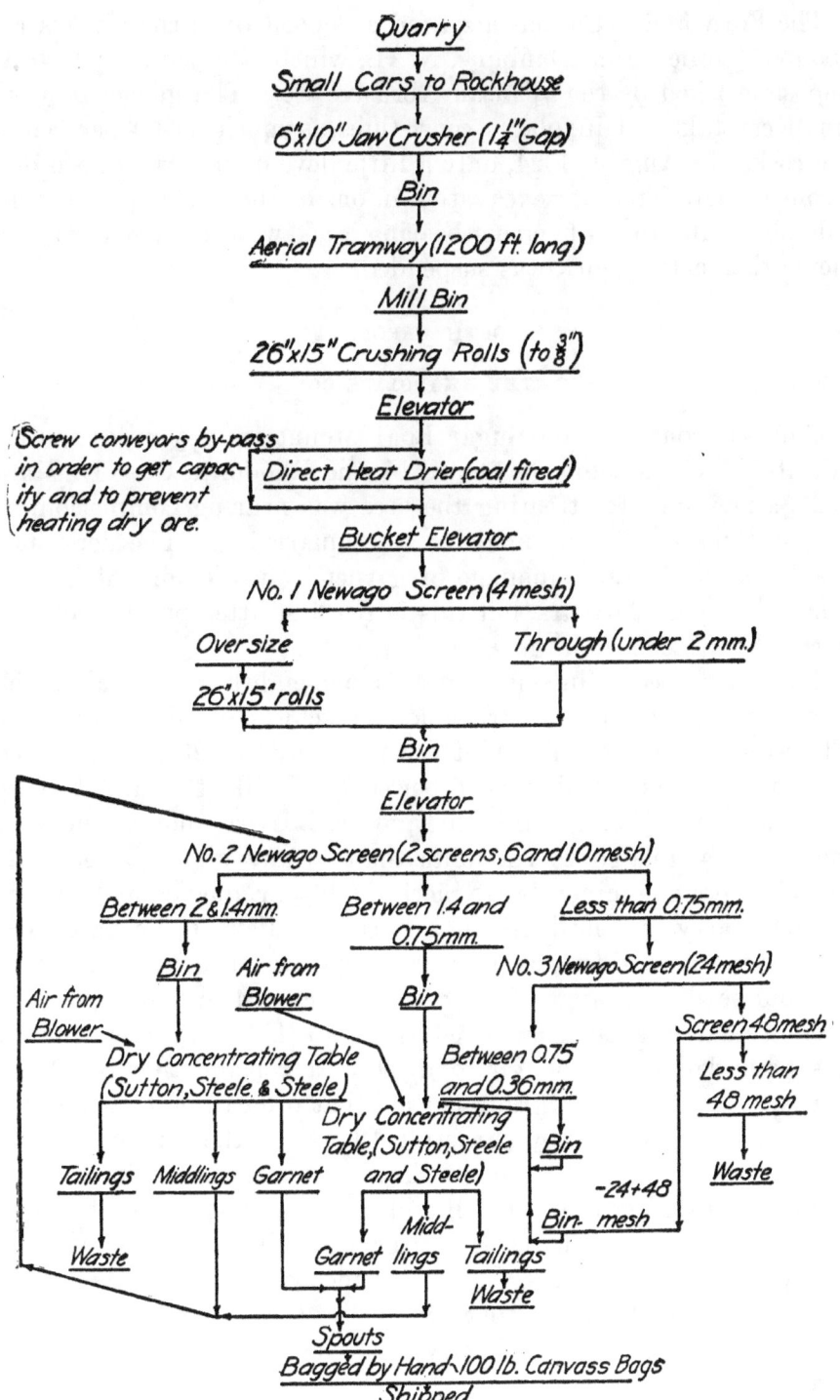

Quarry

Small Cars to Rockhouse

6"×10" Jaw Crusher (1¼" Gap)

Bin

Aerial Tramway (1200 ft. long)

Mill Bin

26"×15" Crushing Rolls (to ⅜")

Elevator

Screw conveyors by-pass in order to get capacity and to prevent heating dry ore.

Direct Heat Drier (coal fired)

Bucket Elevator

No. 1 Newago Screen (4 mesh)

Oversize   Through (under 2 mm.)

26"×15" rolls

Bin

Elevator

No. 2 Newago Screen (2 screens, 6 and 10 mesh)

Between 2 & 1.4 mm.   Between 1.4 and 0.75 mm.   Less than 0.75 mm.

Bin   Air from Blower   No. 3 Newago Screen (24 mesh)

Air from Blower   Bin   Screen 48 mesh

Dry Concentrating Table (Sutton, Steele & Steele)   Between 0.75 and 0.36 mm.   Less than 48 mesh

Dry Concentrating Table, (Sutton, Steele and Steele)   Bin   Waste

Tailings   Middlings   Garnet   −24+48 mesh   Bin

Waste   Garnet   Midd-lings   Tailings

Waste

Spouts

Bagged by Hand 100 lb. Canvass Bags

Shipped

FIGURE 2.—Flow sheet of mill of Wausau Abrasive Co.

FORD MOTOR CO.

The Ford Motor Co. has acquired a deposit of garnetiferous rock about 1½ miles from Danbury, N. H., which is much similar to the deposit worked by the Wausau Abrasive Co. The garnet occurs in small crystals and in places constitutes as much as 60 per cent of the rock. In August, 1924, only a little development had been done. From a small, irregular excavation in one of the outcrops in the hillside about 100 tons of garnet-bearing rock were removed for shipment; then active work was suspended.

NORTH CAROLINA

THE RHODOLITE CO.

Garnetiferous rock on Sugar Loaf Mountain, 2½ miles south of Willits, N. C., is being developed by the Rhodolite Co. In March, 1925, a new mill for treating the ores was nearing completion, and preparations were being made to start quarrying. The deposit has yielded a considerable tonnage of garnet by work done at intervals during the past 25 years, but this is the first attempt to recover the garnet by systematic large-scale operations.

The garnet occurs in small crystals, one-eighth to one-half inch in diameter, in mica and quartz-feldspar schists of unknown extent. The average garnet content of the rock is 20 to 25 per cent; local concentrations are as high as 60 per cent. Unlike the garnet of New York and New Hampshire, this garnet is rose colored and seems more closely related to rhodolite than to any other variety. The principal gangue minerals are biotite, feldspar, quartz, and pyrite.

The quarry and mill are in a narrow valley at the base of the mountain whose sides rise steeply. The overburden, a mixture of soil and bowlders, ranges from a few inches to 4 feet in depth. The present quarry, about a hundred yards up the valley from the mill, has been developed enough to disclose a considerable tonnage of garnet ore. The track grade is such that ore can be transported to the mill in 1-ton cars by gravity. Plans for the mill contemplate dry concentration according to the flow sheet shown in Figure 3. Some small changes in the mill and more equipment to increase the capacity may prove necessary after the mill is in operation. A 24 by 15 inch jaw crusher set at 3.5 inches will crush the ore and will be followed by a gyratory crusher set to 0.75 inch.

The crushed ore will go directly to a drier or be by-passed by a conveyor and elevator to a set of screens. Oversize from the quarter-inch and 12-inch mesh screens will go to two sets of rolls, one to crush to three-eighths inch and the other to 10 mesh. The crushed garnet from these rolls is to be returned to the elevator and re-

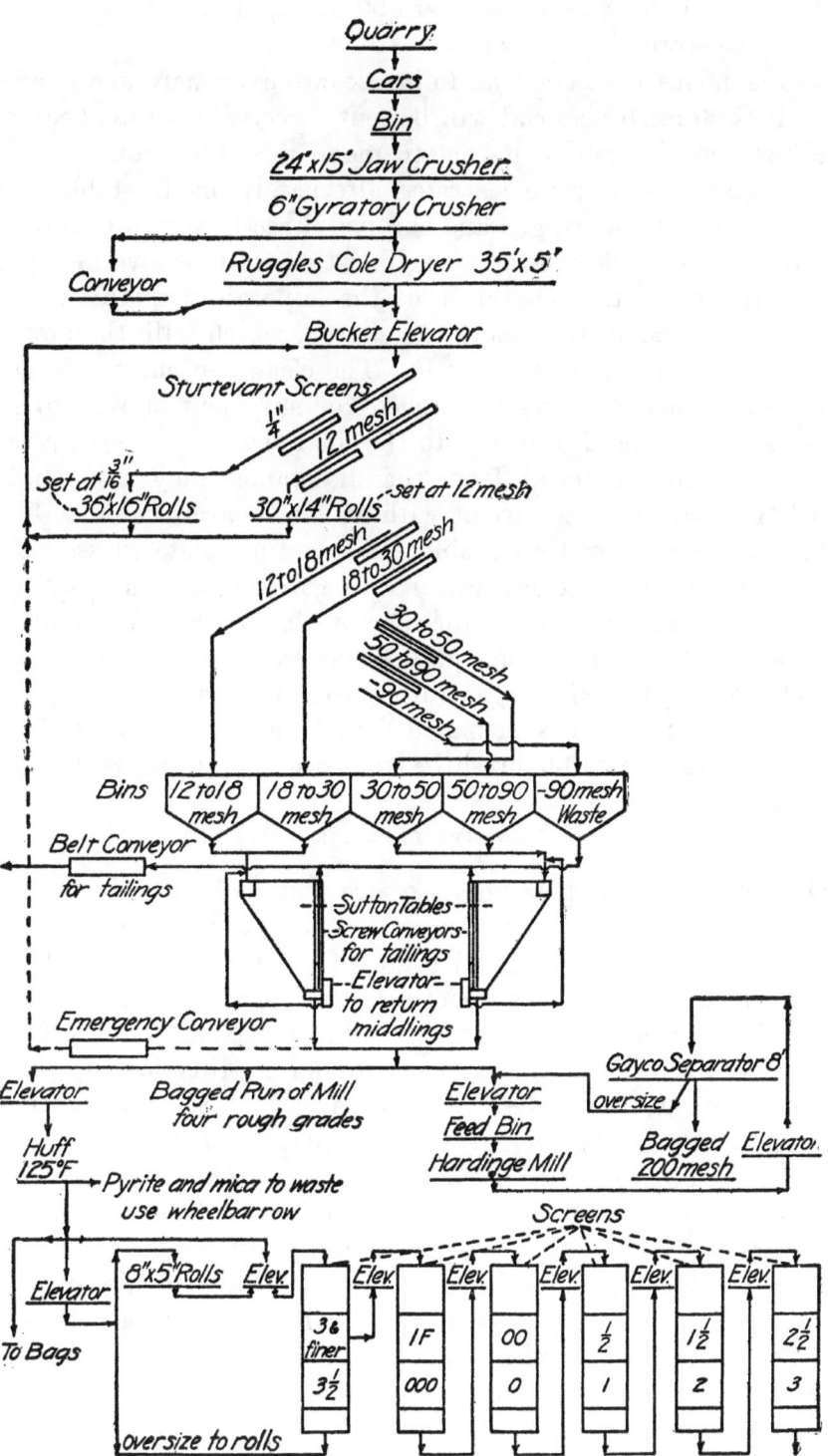

FIGURE 3.—Flow sheet of mill of the Rhodolite Co.

screened.   The final products will be sized as follows: 10 to 18 mesh, 18 to 30 mesh, 30 to 50 mesh, 50 to 90 mesh, minus 90 mesh.   Each size will be stored in a 50-ton bin.

As the garnet does not tend to break into extremely fine particles, the minus 90-mesh material will be sent directly to waste, because it does not contain enough garnet to make recovery profitable.   The sized products will be concentrated alternately on dry tables.   The concentrates may be bagged and shipped directly as run-of-mill garnet or sent to an electrostatic machine that will remove any pyrite and biotite present.   Separation of the minerals with this machine is greatly increased by heating the charge, which will, therefore, be warmed by steam pipes to 125° F.   The clean concentrate from the electrostatic machine may be bagged for shipment or sent to a set of rolls, crushed, and graded into the final sizes by a set of screens.   The run-of-mill material from the dry tables may also go to a Hardinge mill in closed circuit with an air separator that will prepare a 200-mesh powder suitable for grinding plate glass.   If desired, a clean biotite concentrate may be prepared as a by-product.   This type of mica is suitable for concrete facing, rolled roofing, and the manufacturer of ornamental powders.   Two 130-horsepower Diesel-type engines with a generator directly connected will provide power for the mill.   It is estimated that the mill will produce 12 tons of concentrates per 10-hour shift from ore containing 20 to 25 per cent garnet.

OTHER DEPOSITS

Deposits of garnet-bearing rocks are abundant in western North Carolina and a small, irregular production has been reported from several localities.   A large deposit of garnetiferous hornblende-gneiss which outcrops near Shooting Creek in Clay County contains almandite garnet in small crystals seldom exceeding three-fourths inch in diameter.   A black garnet found in this locality is said to be suitable for abrasives.   It is apparently a red almandite containing minute inclusions of black minerals which make it look black on casual examination.   A considerable area of the hornblende-gneiss is exposed on Penland Bald Mountain a few miles from Shooting Creek; it shows garnet in amounts ranging from 2 to 15 per cent. A considerable tonnage of ore that would furnish mill rock containing 5 to 6 per cent of garnet is available.   The garnet has the laminated structure typical of almandite and yields very sharp abrasive grains.   Absence of good roads and the distance to railroad transportation have prevented the development of these deposits on a commercial scale.   As the total amount of garnet-bearing rocks is enormous the district may become an important source of abrasives in the distant future.

Large amounts of almandite garnet suitable for making abrasives occur in the district northwest of Marshall in Madison County. The garnet is present as crystals one-quarter to 1 inch in diameter in a mica schist and the deposits have potential commercial importance.

DEPOSITS IN OTHER STATES

Deposits of garnetiferous rock from which a supply of the mineral suitable for the abrasive market may be obtained are known to occur in Georgia, Virginia, Montana, Colorado, and many western States. The ore reserves of the active producers are large enough to supply the market at the present rate of consumption for many years.

DEPOSITS IN FOREIGN COUNTRIES

Although many deposits of garnet occur outside of the United States, little has been done toward their development. The foreign demand for garnet abrasives has been small and the requirements of the American market could be met most of the time by domestic producers.

SPAIN

For a time some abrasive garnet was produced in the Province of Almeria, Spain, where the garnet occurs in small rounded crystals associated with sand and gravel in stream deposits. This garnet is considered inferior to the standard American article, as it does not make as hard and sharp an abrasive and the small size of the crystals increases the difficulty of obtaining the full range of sizes required for the manufacture of abrasive papers. In the early years of the garnet industry, when the domestic supply was not equal to the demand, imports of Spanish garnet ranged from 500 to a maximum of 3,000 tons annually. In 1924 the imports dropped to approximately 300 tons and at present the domestic production exceeds the demand.

CANADA

Many deposits of garnet have been reported in Canada, some of which could supply a large tonnage of mineral suitable for use as an abrasive. The most important are at Depot Harbor on Parry Island, and in the townships of Ashby, Elzevir, Portland, Loughrin, Dill, and Harcourt in Ontario; at Chegoggin Point, Yarmouth, Nova

Scotia; and in Rawdon and De Ramsey Townships, Quebec. Little commercial development has been done except at two localities.

A few tons of garnet ore have been mined at Depot Harbor, where prospecting disclosed an extensive ore body said to average 15 per cent garnet. Water transportation is available and the erection of a concentrating mill has been considered. The first appreciable output of Canadian garnet was in 1923,[6] when one producer, the Bancroft Mines Syndicate, shipped 1,250 tons of hand-picked ore and concentrates, valued at $100,000, to abrasive paper manufacturers in the United States. The Bancroft Mines Syndicate worked a deposit on Concession 19, lot 9, Ashby Township, Ontario. The garnet (almandite) occurs in small crystals from one-fourth inch to 1 inch in diameter in a hornblende-biotite gneiss and constitutes 25 to 40 per cent of the whole rock mass. An average mill feed of 30 per cent garnet was obtained. The deposit is on a hillside thickly covered with brush and a light overburden of soil, and its extent has not been accurately determined. The quarry has been carried 40 feet into the hillside, leaving a face 20 feet high. Stripping for 150 feet along the strike disclosed uniform ore. The deposit is 15 miles from Bessemer siding, the railroad shipping point. A new road may shorten the wagon haul. The concentrator burned down in the fall of 1923 and little mining has been done since.

INDIA

Garnet is abundant in many of the gneisses and schists of the Indian peninsula and the garnet liberated by disintegration of these rocks has been concentrated in river and sea sands. Gem garnet has been produced in India for centuries and its collection, cutting, and polishing forms a small local industry. Garnetiferous gneisses are common in Orissa and attempts have been made to market a massive garnet rock from the Hazaribagh district. Some garnet recovered from the river sands of the Nellore district has been sold for use as an abrasive. Deposits of garnet, believed to be suitable for abrasive manufacture, are widely distributed in Mysore. The garnet occurs in mica and hornblende schists and gneisses and also as fragments of crystals in stream sands and in the soil. According to a Government report[7] in 1914, more than 1,000 tons of garnet sand for abrasive purposes were collected in the Tinnevelly district of Madras.[7] The Indian production from 1914 to 1919 is reported as follows:

---

[6] Some Canadian nonmetallic minerals; a review of 15 years' progress: Trans., Canadian Inst. Min. and Metal., vol. 28, 1925, pp. 7–10.

[7] Garnet: Bull. Indian Industries and Labours, No. 12, Calcutta, 1921, pp. 49–54.

Production of garnet in India

Year	Garnet sand (hundred-weight)	Value (pounds sterling)	Year	Garnet sand (hundred-weight)	Value (pounds sterling)
1914	[1] 21,440	464	1917	nil.	
1915	[2] 115		1918	nil.	
1916	[2] 470		1919	[3] 1,045	

[1] From Tinnevelly district, Madras.   [3] From Mysore.
[2] From Hyderabad (Deccan), mainly.

Some abrasive garnet has been exported to Great Britain and Europe, but the business was abandoned as unprofitable. The plentiful supply of cheap labor in India should make the cost of production low. The information available does not indicate that Indian garnet is equal to the American article and can compete with it in the manufacture of high-grade abrasives.

NYASALAND

In 1924 samples of a coarsely crystalline garnet from Malawe Hill, Nyasaland, were submitted to manufacturers of abrasives. This garnet was said to be of good quality and suitable for use as an abrasive. Little is known of the extent of the deposit from which the samples were taken or the possibility of producing garnet on a commercial scale.

PRODUCTION OF ABRASIVE GARNET

The production of abrasive garnet in the United States did not attain commercial importance until the North River Garnet Co. began milling in 1893. Since then, except in periods of business depression, the production has indicated a constant annual increase in demand. The table below gives the production of crude garnet in short tons and the value in dollars for the years 1895 to 1924.

Production of abrasive garnet in the United States, 1895–1924.[1]

Year	Short tons	Value	Year	Short tons	Value
1895	3,325	$95,050	1910	3,814	$113,574
1896	2,686	68,877	1911	4,076	121,748
1897	2,554	80,853	1912	4,947	163,237
1898	2,967	86,850	1913	5,308	183,422
1899	2,765	98,325	1914	4,231	145,510
1900	3,185	123,475	1915	4,301	139,584
1901	4,444	158,100	1916	6,171	208,850
1902	3,926	132,820	1917	4,995	198,327
1903	3,950	132,500	1918	4,696	248,161
1904	3,854	117,581	1919	4,944	310,131
1905	5,050	148,095	1920	5,476	434,425
1906	4,650	157,000	1921	3,048	260,687
1907	7,058	211,686	1922	7,054	566,879
1908	1,996	64,620	1923	9,006	688,437
1909	2,972	102,315	1924	8,290	674,176

[1] Figures from Mineral Resources, U. S. Geol. Survey.

MARKETING OF ABRASIVE GARNET

Producers of abrasive garnet customarily ship their product as an unsized concentrate—a mixture of the different sizes of grains produced in milling. A small amount of hand-cobbed ore consisting of irregular fragments of garnet crystals, few exceeding 2 or 3 inches in diameter, is also shipped. The manufacturers of garnet abrasives crush and grade these products to the different sizes necessary for making the finished abrasives. The crude garnet is generally shipped in 100-pound bags and is quoted at a price per net ton f. o. b. shipping point. No extra charge is made for bags. In March, 1925, quotations were as follows: Domestic Adirondack, $85 f. o. b. shipping point; Spanish, $60 c. i. f. port of entry; Canadian, $70 to $80 f. o. b. mines. There is a tendency among producers of garnet to install grading equipment at their mills, and in the future it is likely that the consumers will be able to buy any desired size directly from producers.

To meet the demand of plate-glass makers for a fine grinding powder several producers of garnet have added suitable grinding equipment to their mills. Sized garnet grain is sold by the pound, the price depending on the quality of garnet and the size of grain.

MANUFACTURE OF GARNET ABRASIVES

Over 90 per cent of the garnet produced is made into surface-coated abrasives; the remainder is sold as loose grains or powders used in special grinding and polishing operations.

CRUSHING AND SIZING

The first step in the manufacture of garnet abrasives is to crush and accurately size the crude garnet or concentrate received from the mill operator. Here the tendency of garnet to include other minerals within its crystals becomes evident. In some varieties of garnet this tendency is so strong that the garnet has to be cleaned after each crushing to free it from the foreign minerals liberated. This cleaning is accomplished by reconcentration on a dry table. At some plants it is customary to reconcentrate all garnet concentrates in order to prepare a secondary concentrate of the highest possible purity. Biotite and hornblende, the minerals most commonly associated with garnet, are black, and, although their presence in small amounts does not detract appreciably from the value of an abrasive, their appearance is objectionable, and therefore efforts are made to eliminate them.

The cleaned garnet concentrate is crushed in rolls and screened to different sizes. The grade numbers ordinarily used to indicate the size of garnet range from 20 mesh, known as No. 3½, through Nos. 3, 2½, 2, 1½, 1, ½, 1/0, 2/0, 3/0, 4/0, 5/0, 6/0, and 7/0, which is 220 mesh. A coarser product, Nos. 4 and 5, is prepared by some companies. The accompanying table from a Bureau of Mines report [8] indicates the screen mesh corresponding to the grade number of abrasive garnet and of other common abrasives. In grading garnet extremely accurate sizing within narrow limits is demanded. The

Comparative sizes of abrasive grains [1]

[Figures in columns represent grit numbers commonly used]

Standard screen mesh	Standard size of opening (inches)	Flint paper and cloth	Garnet paper and cloth	Silicon carbide and artificial corundum	Emery and artificial corundum paper and cloth	Silicon carbide and artificial corundum paper disks
			●7/0			
200	.0029					
		●4/0	●6/0	●220		
180	.0033			●200		
		●3/0	●5/0			
160	.0038			●180	●3/0	
		●2/0	●4/0		●2/0	
140	.0042					
			●3/0	●150	●0	
				●120		
120	.0046					
		●0				●120
			●2/0	●100	●100	
100	.0055					●100
90	.0059	●½		●90	●½	●90
			●0		●1	
80	.0068					
		●1		●80		●80
70	.0073			●70	●1½	
						●70
			●½	●60	●2	
60	.0097					●60
		●1½	●1	●50		
50	.0110					
		●2			●2½	
40	.0140		●1½	●40		●46
		●2½	●2	●36	●3	
				●30		●36
30	.0198					
		●3	●2½			
		●3½	●3	●24	●3½	
20	.0340		●3½		●4	●24
			●4			
			●5			
15	.0468					●16
						●12
10	.0650					

[1] In this table the size of grain as related to screen mesh is indicated by the position of the dots between the horizontal lines. Thus, No. 2½ garnet is about 28 mesh.

next coarser number permissible in any one grade is only 5 per cent. Standard grades have been adopted by the American Surface Abrasive Manufacturers, so that the consumer may purchase standardized abrasives from different manufacturers and utilize them with no change or interruption in his operations.

[8] Ladoo, R. B., Garnet: Reports of Investigations, Bureau of Mines, Serial 2347.

Garnet is screened in a series of slightly inclined, rectangular, vibrating frames covered with wire screen-cloth or silk grit gauze. Wire screens are more commonly employed for the coarse sizes and silk gauze for the finer. The silk gauze, which can be obtained in accurate mesh, displays a surprising resistance to the abrasion of a mineral so hard and sharp as garnet. The openings enlarge through wear and the screened products must be constantly tested to insure accurate sizing. The under size from the various screens falls into small bins provided with discharge gates.

COATING THE CLOTH OR PAPER BACKING

Paper or cloth or, sometimes, a combination of the two is coated with this closely sized garnet. Manila fiber, or rope paper, is used for the best quality of garnet papers, and Kraft paper for the cheaper grades. The cloth backings comprise various weights of drills, jeans, or twills. Large rolls of these backings pass continuously through a rotary press which prints the manufacturers' name and the size of the abrasive at regular intervals on the reverse side. Then the backing passes between two rolls, one of which revolves in a trough of glue and spreads a film of the adhesive over the surface, and then passes beneath a hopper from which a shower of sized garnet grains descends upon the glued surface. The excess garnet is shaken off and the cloth or paper is suspended in great loops upon drying racks.

The machinery for handling paper backings is similar to that employed in the manufacture of wall paper. After drying, a second, or "sizing," coat of glue is run over the garnet-coated surface to anchor the grains firmly in place. The best quality of hide glue is employed to coat the paper and anchor the grains. After the garnet paper has dried it is wound in large rolls which later are cut into standard-size sheets (9 by 11 inches) or are fabricated into 50-yard rolls of varying widths or into belts or disks suitable for the various demands of industry.

On a new type, nonclogging, abrasive paper, the garnet grains are scattered by mechanical means so that there is a small space between each grain and those surrounding it. This paper is particularly useful in abrading gummy materials, such as resinous woods, which tend to clog the regular paper by filling the spaces between the closely crowded grains, thereby greatly lessening their abrasive value. Special waterproof sheet abrasives are prepared for certain classes of work which can be done more efficiently when the materials are wet. In the manufacture of these abrasives an adhesive insoluble in water replaces glue.

To produce a satisfactory abrasive requires constant attention and the processes are regulated by laboratory control; the size of the garnet grains, the viscosity of the glue, and the quality of all the raw and finished products are examined repeatedly.

In the manufacture of abrasive papers the cost of the garnet represents 16 to 20 per cent of the total cost; the other items are labor, paper, and glue. The amount of garnet per unit of area depends upon the size of grain, character of backing, strength of glue, and type of paper; nonclogging papers contain less garnet than those that are completely covered with the abrasive.

SUPERIORITY OF GARNET PAPER TO SANDPAPER

Considerable difference of opinion exists concerning the conditions in which the superiority of garnet paper over quartz sandpaper is most marked. For many kinds of work it is universally conceded that the use of garnet paper is more economical than quartz or sandpaper, although its initial cost is greater. In abrading soft, resinous woods garnet and sand papers appear to be nearly equal in efficiency, as both are soon clogged with the abraded particles of wood and the cutting edges so blinded that the paper must be discarded. In working hardwoods and similar substances the superiority of garnet over quartz becomes most pronounced. Reports concerning the effectiveness of garnet on hardwoods vary with local conditions, but indicate that garnet will cut from two to six times as much wood as quartz whether measured by weight or the area of the surface abraded. Garnet does not have enough hardness to abrade the harder metals, but can be used satisfactorily on many of the softer ones, including copper, brass, and bronze. Garnet cuts leather rapidly and evenly.

The bulk of the garnet-coated abrasives is made up into belts, covers for drum sanders, or disks. Belts and drum sanders are driven at high rates of speed, varying from 1,000 to 3,000 feet per minute. A considerable percentage of sheet goods is also consumed by hand use.

A multitude of industries, particularly woodworking, metal working, and leather working, use abrasive garnet.

USES OF GARNET PAPER

WOODWORKING

Garnet paper is employed in woodworking wherever a smooth, natural finish or a prepared surface for varnish or shellac is desired. It is most valuable in the finishing of the highest class of furniture, such as pianos, phonograph and radio cabinets, and home and office furniture.   Manufacture of one piano case is said to require about 25 square feet of garnet paper.[9]   The finishing of the woodwork necessary for the manufacture of wagons, carriages, automobiles, boats, and railroad cars, also uses much garnet paper, as does the smoothing of wheels, spokes, handles for all types of tools, hardwood floors, sash, doors, blinds, and all kinds of millwork.   In fact, garnet paper is of value in nearly every woodworking operation where a smooth surface is desired.

METAL WORKING

Garnet paper for metal working is sold to manufacturers of various machines, automobiles, electric appliances, brass specialties, and metal furniture, who use it for cleaning castings, grinding valves, and finishing surfaces.

LEATHER WORKING

In the manufacture of boots and shoes garnet paper is used for scouring heels and soles.   It is also used in repair shops for finishing the heels and soles to final size and smoothing their edges.

SPECIAL USES

Garnet paper has various special uses.   It is employed to finish rubber, bakelite, and celluloid in much the same manner as in finishing hardwoods.   It has been used as an abrasive in cleaning hardwood and composition floors and other materials by vigorous scouring.   The fine papers are used in finishing felt and silk hats, and also in dental work.   In the weaving of silk, rolls covered with garnet paper are used to draw the fabric from the loom.   Garnet paper is commonly used to remove and rub down paint and varnish on both wood and metal surfaces.

Loose garnet grain for special uses is also sold by the manufacturers of abrasive papers.   In some industries special belts and abrasive shapes have to be made for unusual work that can not be done with the ordinary garnet products.   These special abrasives are prepared by attaching the garnet grains with a suitable adhesive to belts or shapes prepared for the particular requirements.   A small amount of garnet grain is bonded into wheels which have a limited use in the grinding of glass and metals.   The low fusion point of garnet and its susceptibility to alteration by heat make impossible the use of any ceramic bond that would require firing in a kiln.   These wheels are bonded with sodium silicate, rubber, shellac, or similar organic substances.

---

[9] Simonds, H. R., Abrasives play an important part in piano construction: Abrasive Ind., vol. 3, Jan., 1922, pp. 9–13.

GLASS POLISHING

Since the introduction in 1914 of garnet as an abrasive for the grinding of plate glass the demand for this purpose has gradually but constantly increased. As the garnet is used for the finest grinding, it must be prepared as a fine powder, ranging from 200 to 325 mesh. The molten glass is rolled into sheets about one-half inch thick and after annealing is set in plaster of Paris on round steel tables from 24 to 36 feet in diameter which have been machined to perfectly plane surfaces. A table bearing a glass sheet is put under a grinding machine which consists of a revolving spindle that carries and turns the table, glass and all. Two freely revolving iron-shod rollers are suspended above the table; one extends over the center of the table, and the other covers the rest of the diameter. When the table is revolving the rollers are lowered, sand and water are applied between the iron of the rollers and the glass surface, and grinding continues until the glass has a plane surface. The sand must pass a 16 to 20 mesh screen and should not contain any material fine enough to pass 50 mesh. To fine the surface of the glass the grinding is continued with garnet until a satisfactory surface is produced. Then the glass is polished in a similar machine provided with felt-covered rolls using rouge as a polishing medium. After one side is finished the glass is turned over and the process repeated on the other side.

SPECIFICATIONS AND TESTS

There are no definite specifications for abrasive garnet. The only way in which the suitability of a specimen of garnet for abrasive purposes can be determined is by actual use and comparison with other garnet products of accepted value as abrasives. In general the suitability of garnet for use as an abrasive depends on its hardness, toughness, fracture, purity, and the size of the crystals available. The mineral should have a hardness above 7.0 and should be tough enough to withstand considerable shock without shattering, but should break after long continued use and present a new cutting edge rather than wear down to a dull surface. The fracture should be such that sharp cutting edges are constantly produced. The tendency toward conchoidal fracture should not be strong enough to produce thin sharp flakes, which are undesirable.

Crude garnet or garnet concentrate carrying less than 90 per cent of the mineral would hardly be of commercial grade to-day, whereas in the early days of the garnet industry when methods of production were rough a material containing as low as 50 per cent garnet was utilized. The adoption of modern equipment and methods for the concentration of garnet has made possible the marketing of a purer product whose value has been appreciated by consumers. Garnet that occurs in small grains in rock or in loose sand is of inferior value, because the crushed grains tend to have rounded faces and a full range of commercial sizes can not be obtained. Color is of no value as an indicator of abrasive qualities.

The determination of the hardness of any mineral is at best an inaccurate procedure. No method is known by which the hardness may be positively determined, and it is, therefore, necessary to resort to a scale by which the hardness of any mineral may be expressed in a relative manner. Moh's scale consists of 10 common minerals arranged in order of increasing hardness. On this scale feldspar is 6, quartz 7, topaz 8, corundum 9, and diamond 10. Abrasive garnet scratches quartz, but in turn is scratched by topaz, and therefore possesses a hardness between these limits which is commonly expressed as 7.5. Some specimens of garnet sometimes appear to have a hardness closely approaching 8.

CHEMICAL ANALYSIS

Chemical analysis is of little value in determining the suitability of a garnet for use as an abrasive, because abrasiveness depends on physical properties. To obtain accurate information on the composition of any specimen of garnet the sample must be prepared with great care. All foreign minerals must be removed, and as these are often present in very small particles it becomes necessary to crush the garnet, examine the powder with a microscope, and remove all impurities. The separation of garnet from other minerals may sometimes be expedited by the use of heavy solutions. Samples from five of the most important commercial sources were analyzed. As analyses of samples taken from the same deposit often show considerable difference, the analyses below may not represent absolutely the garnet in the localities named, but are indicative of the general type of garnet.

The five samples represent the following localities: 1, Massive garnet from the Adirondacks, New York; 2, crystal garnet from the Adirondacks, New York; 3, "rhodolite" garnet from western North Carolina; 4, imported Spanish garnet; 5, New Hampshire garnet.

Analyses of garnet

	1 Adirondack, massive [1]	2 Adirondack, crystal [1]	3 North Carolina [1]	4 Imported Spanish [2]	5 New Hampshire [2]
Silica (SiO_2)	38.92	40.24	38.52	37.06	37.39
Alumina (Al_2O_3)	22.77	20.06	21.53	26.92	20.46
Iron oxide (ferric) (Fe_2O_3)	1.80	4.65	2.72	0.00	2.89
Iron oxide (ferrous) (FeO)	23.34	18.58	27.75	32.24	31.87
Calcium oxide (CaO)	5.22	5.34	2.03	1.02	.92
Magnesium oxide (MgO)	7.09	11.18	8.28	1.86	2.46
Manganese (MnO)	.15	.25	.12	2.93	3.47
Total	99.29	100.30	100.95	102.03	99.46

[1] Analyses by E. E. Berger.   [2] Analyses by M. Farnsworth.

TEST OF ABRASIVENESS

To test the efficiency of abrasive garnet in the form of loose grain is difficult and unsatisfactory, therefore testing is done after the sized grain has been fixed to the cloth or paper backing. The usual test is abrading a standard wooden block for a given length of time and determining the amount of material removed. The test procedure differs. The paper may be made up into an endless belt or a disk, or applied to a small drum sander. To obtain uniform material the blocks of wood are cut from one plank. Some operators prefer to use one block of wood and change the abrasive working on it in order to be assured of identical conditions. The test commonly lasts 10 minutes and the amount of wood abraded in this time is determined by weighing or the loss may be determined every minute and the results plotted in a curve. A standard abrasive paper of known efficiency is used as an index and the results of an unknown paper are reported as a percentage of this standard.

FUTURE DEVELOPMENT OF THE INDUSTRY

The present condition of the industry indicates that the utilization of garnet may be expected to increase. In the year 1924 the industry made its greatest expansion with the erection of new mills in New York and North Carolina. The potential output of the mills now in construction may be estimated to be in excess of 20,000 tons annually. As the annual consumption in 1924 was about 9,000 tons, it is natural to expect that attempts will be made to broaden the market and find an outlet for the excess production.

Some of the present markets for garnet can be expanded considerably. The use of garnet in the plate-glass industry is limited. If the practice of fine grinding with garnet should become universal in the plate-glass industry an additional demand for several thousand tons annually is possible. The use of abrasive garnet is peculiarly American and as its use in other countries is small the export business has been practically negligible. The introduction of garnet products into the industries of foreign countries, particularly the wood, metal, and leather working trades of Europe, should provide an outlet for a considerable tonnage.

The possibility of finding new uses for garnet has been the subject of some investigation. Garnet grains have been used for facing tile and concrete block; besides being ornamental the garnet presents a surface having great resistance to wear. Substitution of garnet sand for quartz sand in sand blasting has been advocated, but garnet costs considerably more than quartz sand; and hence could be used only where its superior abrasive qualities are of value and the used grain can be recovered.

A possibly large market exists in the surfacing and polishing of the softer ornamental stones, such as marble, slate and serpentine. Large amounts of these stones are cut annually into thin slabs which are used for floors, wall panels, and architectural work. The blocks of stone are sawed into slabs with gang saws—strips of steel fed with sand. The actual cutting is done by the sand being dragged against the marble surface by the steel blade. According to Sewell,[10] if an abrasive could be made available that would cut three or more times as fast as sand, and not cost more than 10 times as much per ton— or if its lasting qualities were proportional to its higher price if it were very expensive—the cutting of marble would be greatly facilitated. The possibility of garnet meeting these requirements will have to be determined by actual use.

Plate III shows the shape of the grains in crushed garnet.

MILLING OF GARNET

The chief factors that control the behavior of garnet during concentration are the identity and characteristics of the associated minerals, the size of the garnet crystals, and the percentage of them in the ore. The common associates of garnet are feldspar, mica, hornblende, pyroxene, and quartz. Magnetite, pyrite, ilmenite, limonite, pyrrhotite, and occasionally chalcopyrite, rutile, zoisite, and corundum are associated in much smaller percentages. The sodium feldspar, plagioclase, is a more common associate of garnet than the potash-bearing varieties. Biotite, the black iron-bearing mica, is seen more often than muscovite, the white mica. Hypersthene and diopside are the most common varieties of pyroxene. The most important characteristic controlling the behavior of these minerals during concentration is specific gravity.

SPECIFIC GRAVITY OF GARNET AND OF ASSOCIATED MINERALS

The authors determined the specific gravity of a large number of mineral specimens from two of the most important deposits actively worked in the Adirondack region of New York State. These specimens, here classified as clean garnet, clean hornblende, locked garnet, and common gangue, were collected from the feed, concentrate, and tailings of the mills, and from representative localities in the quarries. The clean garnet represents selected specimens in which no foreign minerals were visible, and the clean hornblende represents similarly selected specimens. The locked garnet represents specimens in which the garnet in varying amounts is attached to the gangue minerals. The common gangue is the mixture of minerals that composes the rock surrounding the garnet.

---

[10] Sewell, J. S., Handbook of marble, chapter 3: Stone, Feb., 1925, pp. 36–41.

The mineralization of these deposits is comparatively simple. In addition to garnet the rocks are composed essentially of feldspar, hornblende (occasionally present in pure, fairly large masses), pyroxene, and a little biotite. These minerals form about 98 per cent of the rocks which are fairly fine textured. Magnetite, pyrite, and ilmenite are disseminated in such small particles that they have little effect on the specific gravity of the rock.

Specific gravity of garnet, hornblende, and gangue

SPECIMENS FROM DEPOSIT 1

	Maximum specific gravity	Minimum specific gravity	Average of all determinations
Clean garnet	4.11	3.88	3.95
Clean hornblende	3.24	3.07	3.16
Locked garnet	3.69	3.31	3.50
Common gangue	3.18	2.96	3.06

SPECIMENS FROM DEPOSIT 2

Clean garnet	4.19	3.90	4.00
Clean hornblende	3.19	3.10	3.16
Locked garnet	3.44	3.00	3.23
Common gangue	2.93	2.68	2.79

The specific gravity of the clean hornblende is of interest because of the common reports that the gravity of the hornblende and garnet lie so close together that separation of the two minerals by gravity concentration is very difficult. The specific gravity of the locked garnet has a considerable range, as might be expected. Garnet usually tends to break clean from the surrounding minerals when they are crushed, but these minerals sometimes adhere more strongly and the crushed ore contains fragments in which the garnet is fast to the gangue. These fragments are most conspicuous after the primary crushing. The secondary crushing is so regulated as to free the garnet completely.

Biotite and muscovite mica and quartz are not commonly associated with garnet in the Adirondack district but may be in other localities. The table below gives the specific gravity of the common accessory minerals:

Specific gravity of minerals associated with garnet

	Specific gravity		Specific gravity
Biotite	2.70–3.20	Corundum	3.90–4.10
Muscovite	2.80–3.10	Limonite	3.40–4.00
Quartz	2.65	Magnetite	4.90–5.20
Feldspar	2.60–2.76	Pyrite	4.90–5.20
Pyroxene	3.20–3.60	Pyrrhotite	4.50–4.60
Rutile	4.20–4.30	Chalcopyrite	4.10–4.30

Presence or absence of accessory minerals has little effect on the main operations in the concentration of garnet but may necessitate adjustments of minor details. The larger the garnet crystals the greater the proportion of clean garnet liberated on primary crushing. The tendency of garnet to include other minerals gives trouble because these minerals are often disseminated throughout the garnet crystals in such small particles as to require very fine crushing to insure a clean garnet product.

The percentage of garnet in commercial ores, as indicated in the discussion of milling operations, varies greatly—from a minimum of 6 per cent to a maximum of 60 per cent. The necessity of adjusting the flow sheet of any mill to the richness of the feed is evident.

ASSAY OF GARNET ORES

MICROSCOPIC EXAMINATION

Examination of mill products shows at once that a method of analysis by which the proportion of garnet can be determined accurately will be a great aid. The determination of garnet is a difficult operation and one which does not permit a high degree of accuracy. Because of the complex composition of garnet and the minerals with which it is commonly associated the proportion of garnet can not be determined by chemical analysis, and resort to mechanical means is necessary. Before the garnet can be separated from the associated minerals the ore must be crushed finely enough to liberate the garnet. Then a weighed portion of the crushed ore may be picked by hand with the aid of a small pair of tweezers and a magnifying glass, and the collected particles of garnet weighed, and the garnet content of the ore or mill product calculated. This method is laborious, and careful manipulation is necessary to get accurate results.

USE OF HEAVY SOLUTIONS

The assay of garnet ores may be facilitated by the use of heavy solutions. Although the specific gravity of the garnet group ranges from 3.4 to 4.3, the varieties of garnet of industrial importance have much closer limits, generally from 3.9 to 4. Except in isolated cases of unusual mineralization the gravity of the associated minerals does not exceed 3.20. It is obvious that the gangue minerals would float on a liquid having a specific gravity between 3.20 and 3.90, although the garnet would sink. With the aid of such liquids the garnet content of a sample of crushed ore may be determined.

Few liquids are suitable. One of the most convenient is methylene iodide, which has a gravity of 3.3. It is miscible in all proportions with benzol, gravity 0.88, and thus liquids of lower gravity can be prepared. Klein's solution (specific gravity 3.28), a solution of cadmium boro-tungstate in water, may also be employed. As the preparation of heavy solutions is rather costly, their use has been limited. Accurate methods of sampling and assay have seldom been employed in the garnet industry. The garnet content of a deposit has been determined either by the estimate, based on careful examination, of an experienced man or by milling and concentrating a small experimental tonnage. The development of cheaply prepared heavy solutions and a simple technique for their use would be of considerable value to the operator of a garnet mill by enabling him to assay the various products from the concentrating equipment and to check up the efficiency of the mill.

At its Mississippi Valley station the Bureau of Mines is attempting to develop various heavy solutions that will be suitable for such determinations as that of garnet in its ores. Chemical analyses are more or less valueless in determining the content of many non-metallic minerals in their ores, because many of the minerals have a complex composition, and usually every element they contain is present in some of the associated minerals. Thus an assay for iron in a garnet ore gives no accurate measure of the amount of garnet present because the associated hornblende, magnetite, and mica also contain iron. Resort may be had to physical analysis by the use of heavy solutions. Mineralogists and geologists have used these solutions for isolating and identifying minerals, but the metallurgist has been backward in adopting them for ore dressing. As most of the concentration processes depend primarily on specific gravity, physical fractionation by heavy solutions provides a most effective tool in dissecting the structure and character of an ore with respect to suitable methods of concentration. For any ore that is to be concentrated by methods depending on specific gravity, physical dissection by means of these solutions provides a most valuable means for determining the nature of the mineral which is not recoverable by a particular machine. The character and the nature of the mineral lost in any tailing should be accurately known (heavy solutions assist greatly in this determination) before the mill operator can intelligently decide whether this mineral is recoverable in its present condition.

HEAVY SOLUTIONS AVAILABLE

For the study of garnet ores several heavy solutions are available. For removing such minerals as quartz, feldspar, etc., acetylene tetrabromide, which when chemically pure has a specific gravity of about 2.95, can be used; it is liquid at room temperature and may be diluted with carbon tetrachloride, whose specific gravity is about 1.60. For isolating the hornblende and minerals of similar specific gravity the product which sinks in the acetylene tetrabromide solution could be tested with a solution of tin tetrabromide (specific gravity 3.31 when the pure salt is melted), which gives a clear liquid when melted at 30° C., slightly above room temperature. It may be diluted with carbon tetrachloride to obtain lower specific gravities. If the sink product in the tin bromide is now subjected to an antimony tribromide solution (specific gravity 3.7, liquid at 94° C.), the float product will be particles composed of garnet locked to the associated minerals, such as hornblende, and the sink product will be primarily free grains of garnet with small amounts of such minerals as magnetite, ilmenite, rutile, etc. To separate these minerals from the garnet the sink product in the antimony bromide may be put in a solution of thallium-silver nitrate (specific gravity about 4.8). This substance, made from the nitrates of the two metals, melts in the range of 65° to 80° C., the melting point varying with the relative amounts of the two nitrates present, is miscible with water, and a liquid of practically any specific gravity between 1.0 and 4.8 may be obtained. Water solutions of this double nitrate might be used for all of the separations mentioned above, but the substance attacks sulphides liberation of metallic silver. For a discussion of various heavy solutions, see articles by Retgers.[11]

SUGGESTED PROCEDURE IN FRACTIONATION

The above discussion indicates briefly how a physical fractionation of a garnet ore might be made. To determine the percentage of garnet the following procedure might be followed: Garnet like that in the New York deposits is practically free from all associated minerals at 35 mesh; furthermore, garnet slimes very little. The sample to be tested for garnet content could be crushed carefully through 35 mesh. Then 5 to 10 grams or even more of the crushed sample could be stirred gently in a beaker of water and the genuine mud or slime could be decanted off, because the presence of mud or genuine slime finer than 325 mesh gives trouble in a separation with heavy solutions. Next the sample should be dried thoroughly and tested with the antimony tribromide solution (specific gravity 3.7) or a water solution of thallium-silver nitrate of about the same gravity. The garnet and the small amounts of such minerals as magnetite, ilmenite, and rutile would sink. If more than very slight quantities of these minerals were present with the garnet, they might be separated by using a solution of thallium-silver nitrate of higher gravity. This procedure would give a fairly accurate determination

[11] Retgers, J. W. [Concerning gravity solutions for the separation of minerals]: Neues Jahrb. Mineral., Bd. 2, 1889. [Experiments to obtain new gravity solutions for the separation of minerals]: Neues Jahrb. Mineral., Bd. 2, 1896.

of the garnet content. Under any circumstances it is probable that the amount of garnet removed by the preliminary treatment to remove mud would be negligible. Various details must be perfected before such a method of physical analysis will rank in extreme accuracy with the methods of chemical analysis used for other minerals, but it will be an advance over the rule-of-thumb methods, in which one judges the garnet content by eye or by tedious hand sorting.

PRINCIPLES OF GARNET MILLING

The basic principles of milling garnet have been illustrated in the descriptions of plant operations. Mills may use wet or dry concentration.

WET MILLING

In wet milling the general procedure is to subject the ore to a primary crushing that liberates most of the garnet from the gangue minerals, screen the crushed material, and concentrate the coarse-size products on jigs which produce a concentrate, middling, and tailing. The concentrate and tailing may be withdrawn immediately from the mill circulation as finished products or a dirty concentrate may be made which must be cleaned by further concentration. The middling is sent to secondary crushers and broken again to liberate its garnet content. The crushed middling product rejoins the mill circulation and is sized and concentrated. The finely crushed ore that has been separated by the screens is treated on tables or by special machines that have been designed for this particular work.

DRY MILLING

In dry milling the ore passes through a primary crusher that discharges into a direct-heat drier. The dried product is screened, the oversize crushed further, and returned to the screen. The undersize, which has been separated into fractions having the desired dimensions, is stored in bins. The sized products from these bins are concentrated on dry tables which make a concentrate, middling, and tailing. The concentrate and tailing are finished products, but the middling is returned to the mill circulation for further treatment.

It is obvious that the amount of crushing to be done by the primary crushers, the sizes of screens for sizing the ore, and the type of equipment for concentration all depend so largely on local conditions that no standard technic for the concentration of garnet has been established. The peculiarities in mineralization of each deposit must be studied and a concentrating process devised that will recover the largest possible percentage of the garnet at the lowest possible cost.

CLASSES OF ORE

Garnet ores may be classified as of two types. In the first the garnet occurs in large crystals, but the percentage of garnet in the ore is comparatively small. A representative ore of this type would contain crystals having a diameter of three-fourths inch or more which would form 4 to 15 per cent of the total rock mass. This type of ore is characteristic of the Adirondack section of New York. In the second type the mineral occurs in small crystals, seldom exceeding three-eighths inch in diameter. In some localities these crystals occur in such profusion that the garnet content of the rock may be as great as 60 per cent. This type is characteristic of the garnet deposits of New Hampshire and North Carolina.

COMMENTS ON CRUSHING PRACTICE

To obtain some information concerning the behavior of garnet during concentration the authors studied the performance of one mill. This study furnished data from which generalized conclusions were drawn as follows:

The extent to which primary crushing should be carried before concentration is begun depends upon the size of the garnet crystals and the facility with which the garnet breaks from the other minerals. Rough concentration of some of the New York ores could begin on material as coarse as three-fourths inch, because many of the crystals are larger than this and preliminary crushing liberates much of the garnet. The ores of New Hampshire and North Carolina, in which the garnet crystals are much smaller, would require much finer crushing. Many engineers allow the consideration "At what size is the economic mineral completely free from gangue?" to govern the size at which concentration shall begin. The antithesis of this statement—that is, "When is a goodly portion of the gangue free from the economic minerals?"—is not often considered. The two principles may appear to have the same meaning, but they do not.

Often it may be found that at some particular size as much as 75 per cent of the gangue is barren of the economic mineral but only a small percentage of the economic mineral is free from gangue. Then an effort should certainly be made to begin concentration at that size, in order that as much of the barren gangue as possible be removed from the mill circulation and immediately sent to the tailing pile, thereby increasing the capacity of the mill and decreasing costs by the elimination of worthless material which no further treatment can make of value. Such a consideration should govern the size of initial grinding at which concentration should begin. The primary concentration might not give a finished concentrate, but it should give a finished tailing whose production would be of most importance. Too much stress is often placed on getting a concentrate as quickly as possible, whereas the production of a "quick" tailing should be emphasized, especially if the gangue minerals comprise the bulk of the ore. Consideration should be given to the idea "Mill to produce finished tailing and the concentrate will take care of itself."

The use of heavy solutions is a valuable aid in determining the point at which most of the gangue is free from the economic mineral. One can hardly expect to recover a clean concentrate while making a clean tailing, and the dirty concentrate made will require further treatment for separation into a clean concentrate and a middling. As this middling is the part of the ore in which the garnet and the gangue minerals are still mechanically locked, it will require further grinding. In most ores the garnet is not liberated completely until ground to about 35 mesh, and for some ores the grinding must continue to 90 mesh to permit the recovery of all the garnet.

If such coarse crushing as one-fourth inch to three-fourths inch be practiced, to "bull-jig" at least the coarser part of the mill feed would be advisable as an attempt to remove as much tonnage as possible from the mill system by producing a clean tailing. The other products from the bull-jig would receive further treatment, such as regrinding and further concentration. The great advantage of removing as much tailing as possible in this first concentration is that a smaller tonnage has to be handled in subsequent milling. The tendency of laminated garnet to break in large flat flakes requires finer grinding than would be necessary otherwise as the large surface of these flakes prevents the garnet from sinking to the concentrate level in the jigs. Garnet in this form tends to float with the tailing and may be lost with it unless the crushing is fine enough to break down the flat particles. If much middling is produced in this first concentration especial care must be taken to provide effective devices and methods for recovering it; a fixed-sieve jig possibly has some advantage over a movable-sieve jig. Probably the best method is to try to draw these middlings at some level above the sieve level. This drawing is more difficult on a movable-sieve jig than on a fixed-sieve jig.

Fortunately, garnet does not slime during the crushing of the ore, consequently the treatment of slime, a serious problem in the milling of many zinc, lead, copper, and gold ores, does not have to be considered. This absence of slime is also fortunate with respect to utilization, as the commercial demands are for comparatively coarse material. An accurate sample of the feed to one mill showed the following screen analysis:

Screen analysis of mill feed

Screened product, size	Per cent by weight	Garnet content, per cent	Distribution of garnet, per cent
−¾-inch+½-inch	14.37	5.25	6.77
−½-inch+⅜-inch	12.81	6.60	7.59
−⅜-inch+¼-inch	16.11	11.05	11.94
−¼-inch+4-mesh	5.42	15.20	7.40
−4-mesh+8-mesh	15.21	19.45	26.56
−8-mesh+10-mesh	6.89	21.40	13.24
−10-mesh+20-mesh	9.44	17.60	14.92
−20-mesh+28-mesh	3.54	11.00	3.49
−28-mesh+35-mesh	4.12	10.20	3.77
−35-mesh+48-mesh	2.76	9.40	2.33
−48-mesh+65-mesh	2.34	5.00	1.05
−65-mesh+100-mesh	1.87	2.50	.42
−100-mesh+150-mesh	1.36	2.00	.24
−150-mesh+250-mesh	1.38	1.30	.16
−250-mesh+325-mesh	.52	.60	.03
−325-mesh	1.86	.50	.08
Composite mill feed	100.0	11.14	100.00

The third column in the table gives the garnet content or assay as determined by hand sorting and weighing the products obtained or by microscopic count, and the fourth column indicates the percentage of the total garnet content in each size.

The quantity of garnet in material finer than 100 mesh in the mill feed is only 0.52 per cent of the total of all sizes. This figure—obtained by adding the percentages of garnet present in the sizes from minus 100 plus 150 mesh to minus 325 mesh in the column "Distribution of the garnet"—indicates that this material contains very little garnet fine enough to be considered slime. This mill feed had been prepared by crushing the run-of-mine rock from the quarry in a jaw crusher and screening in a three-fourth inch round-hole trommel. The oversize from the trommel was recrushed in a gyratory crusher and returned to the trommel for further screening, so that the entire mill feed was finally reduced to this size. Ores requiring a finer preliminary crushing would naturally contain larger quantities of fine garnet but it is improbable that any true slime would appear.

The middlings produced by the primary concentration must be crushed in order to free the locked garnet before concentration can be continued. If the mill feed were crushed initially through three-fourths inch and the portion between three-fourths inch and one-half inch were bull-jigged to remove some finished tailing, the middling concentrate from this operation should be crushed through one-half inch and this material should be added to the minus one-half-inch portion of the original mill feed to give a product suitable for further concentration by jigging.

Two procedures are open in jigging. The feed may be screened and separated into sized fractions which are treated separately or the unsized material produced by crushing to a definite limit may be concentrated immediately without further sizing. Jigging a sized feed is the method followed in the New York mills. The reason given for this sizing is that the specific gravities of the garnet and the associated hornblende are not far enough apart to enable a separation without sizing. But as already shown the difference in gravities is about 0.8 and it is not at all impossible that the jigging of unsized feed as practiced in the Wisconsin and Tri-State zinc districts could be developed to the point where it would be successful with garnet ores. Flat flakes of laminated garnet might seriously interfere with jigging an unsized feed but if this jigging could be practiced it should appreciably decrease the cost of milling.

No definite limit may be established as to the size at which jigging must stop and other methods be utilized to recover the fine material, but it is probable that anything finer than 2 millimeters should be concentrated on tables or special devices. In wet concentration accurate sizing before tabling is unnecessary although some simple classification should be practiced. Division into a " coarse-sand " product and a " fine-sand " product would probably suffice for satisfactory tabling. Because it contains no garnet, the extremely fine material could be sent directly to waste from the classifier. The heavy accessory minerals, such as pyrite, magnetite, and ilmenite, are in such small crystals that they are not liberated until crushed very fine, and commonly appear on the tables in a small streak above the garnet. It is probable that a moderate classification before tabling would largely eliminate these heavy minerals.

The concentration of garnet in dry mills on pneumatic tables leads either to a mill of small capacity or to a large investment. Wet milling is to be preferred because it is more efficient and economical. Dry milling relies on tables that have a capacity much less than the jigs of the wet mill. In a dry mill there can be little variety of treatment, as all the concentration must be done on one type of machine. The dust losses are high and the dust makes conditions unpleasant for the workmen and causes excessive wear of equipment.

A peculiar and local condition was noted in one garnet mill. Wet milling caused a reaction in the crushed ore whereby a reagent was produced which attacked the garnet. This reagent was believed to be sulphurous acid or ferric sulphate formed from the pyrite in the ore. Garnet concentrated by wet milling showed considerable deterioration after storage for a few weeks. Although this deterioration did not seriously injure the abrasive qualities of the garnet it caused a very objectionable appearance which impaired the value. Dry milling was therefore adopted.

There seems to be no reason why the methods ordinarily used for dressing metallic ores can not be applied in their entirety to such nonmetallic ores as those of garnet. Some garnet mills furnish good examples of such applications. It is to be expected that in future the technology of garnet will show greater refinements in concentration and a more economical recovery of the mineral.

MANUFACTURERS OF ABRASIVE GARNET PRODUCTS

The names and addresses of manufacturers of garnet products in the United States and Canada are listed below:

UNITED STATES

American Glue Co., East Walpole, Mass.
Armour Sandpaper Works, Chicago, Ill.
H. H. Barton & Son Co., Philadelphia, Pa.
Baeder-Adamson Co., Philadelphia, Pa.
Herman Behr & Co. (Inc.), Brooklyn, N. Y.
The Carborundum Co., Niagara Falls, N. Y.
Manning Abrasive Co., Troy, N. Y.
Minnesota Mining & Manufacturing Co., St. Paul, Minn.
United States Sand Paper Co., Williamsport, Pa.
Wausau Abrasive Co., Wausau, Wis.

CANADA

Western Abrasives (Ltd.), Victoria, British Columbia.
Abrasive (Ltd.), Brantford, Ontario.

SELECTED BIBLIOGRAPHY

Many references to the use of garnet as a semiprecious stone have been published; literature on its far more important use as an abrasive is less plentiful. Below are listed the more important articles on the occurrence and utilization of abrasive garnet.

GENERAL

MINERAL RESOURCES OF THE UNITED STATES: Chapter on abrasives, U. S. Geol. Survey, annual, particularly volumes for 1911 and 1913.

MINERAL INDUSTRY: Chapter on abrasives, annual, particularly volumes for 1894, 1898, 1918, 1919, 1923.

ABRASIVE INDUSTRY: Cleveland, Ohio, monthly journal.

OCCURRENCE, MINING AND MILLING

BARDWELL, EARL S., Garnet milling in New Hampshire: Eng. and Min. Jour.-Press, vol. 91, June 17, 1911, pp. 1209–1210.

BOWMAN, FRANCIS D., Wet sanding practice gains in favor: Abrasive Ind., vol. 5, June, 1924, p. 147.

EARDLEY-WILMOT, V. L., Increased demand and use for garnets: Canadian Min. Jour., July 18, 1924, pp. 685–686.

GLASS INDUSTRY, The use of garnet for glass grinding: Vol. 6, March, 1925, pp. 51–52.

GREGORY, H. E., Garnet deposits of the Navajo Reservation, Arizona and Utah: Econ. Geol., vol. 11, April–May, 1916, pp. 223–230.

HOOPER, F. C., The American garnet industry: Min. Ind., vol. 6, pp. 20–26, 1897.

KATZ, FRANK, Garnet in North Carolina and the market for abrasive garnet: Eng. and Min. Jour., vol. 107, May 24, 1919, pp. 903–906.

LADOO, R. B., Garnet: Report of Investigations, U. S. Bureau of Mines, Serial No. 2347.

MENNIE, T. S., Modern garnet mills operated: Abrasive Ind., vol. 6, February, 1925, pp. 51–54.

MILLER, W. T., The garnet deposits of Warren County, N. Y.: Econ. Geol., vol. 7, August, 1912, pp. 493–501.

MILLER, W. T., Geology of the North Creek quadrangle, Warren County, N. Y.: New York State Mus. Bull. 170, 1914, pp. 78–82.

NEWLAND, D. H., Mineral resources of the State of New York: New York State Mus. Bull. 223–224, 1921, pp. 79–85.

POWER, HENRY R., Preparing garnet abrasives: Eng. and Min. Jour.-Press, vol. 118, Oct. 4, 1924, p. 532.

RICHARDS, LOUIS M., Garnet deposits of Georgia: Min. World, vol. 34, June 3, 1911, p. 1135.

SKERRETT, ROBERT G., The world's greatest garnet quarry: Compressed Air Mag., vol. 28, August, 1923, pp. 81–585.

WATSON, T. L., Mineral resources of Virginia: The Virginia-Jamestown Exposition Comm., 1907, pp. 287–289.

WORMSER, F. E., Mining, concentrating, and marketing garnet: Eng. and Min. Jour.-Press, vol. 118, Oct. 4, 1924, pp. 525–531.

INDEX

O

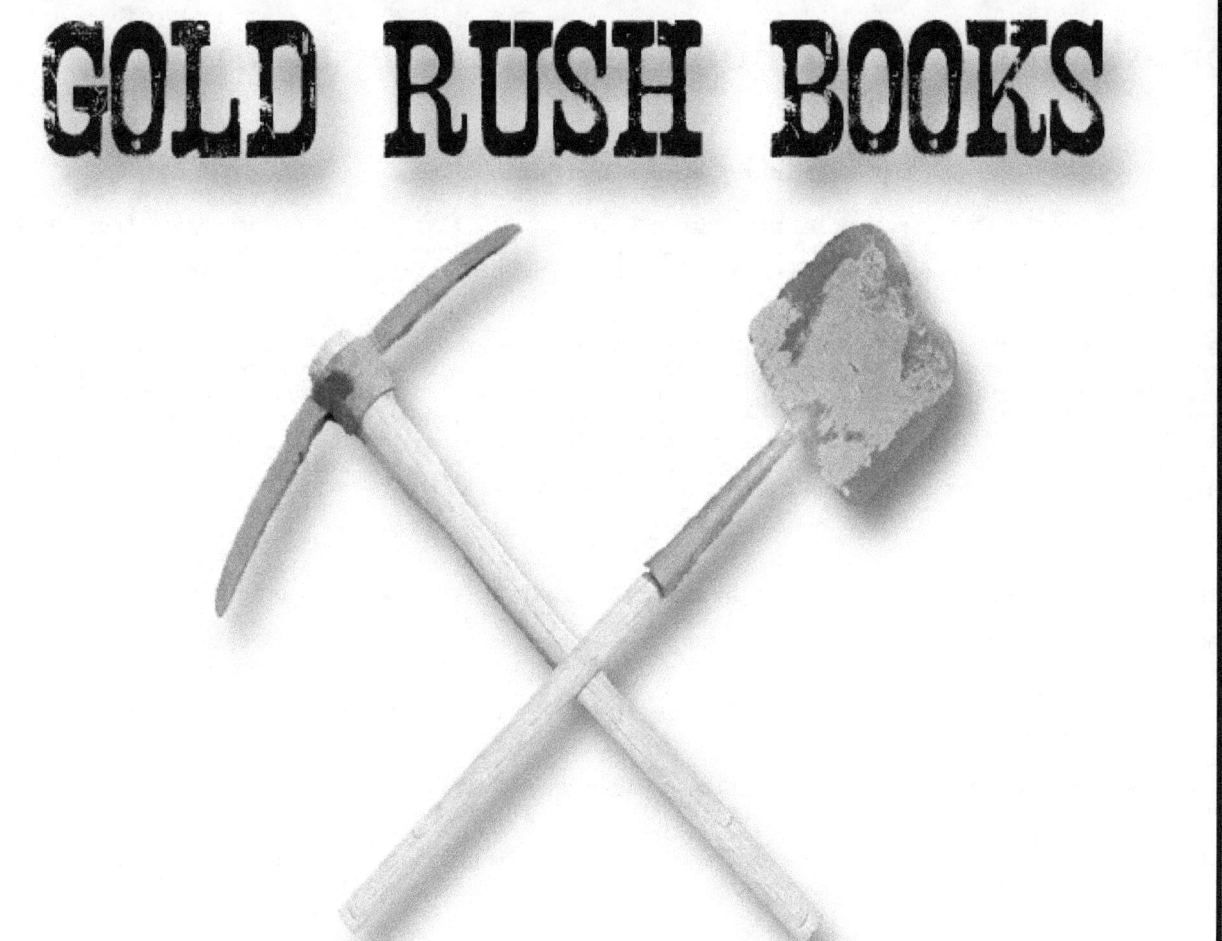

Books On Mining

Visit: www.goldminingbooks.com to order your copies or ask your favorite book seller to offer them.

Mining Books by Kerby Jackson

<u>Gold Dust: Stories From Oregon's Mining Years</u> - Oregon mining historian and prospector, Kerby Jackson, brings you a treasure trove of seventeen stories on Southern Oregon's rich history of gold prospecting, the prospectors and their discoveries, and the breathtaking areas they settled in and made homes. 5″ X 8″, 98 ppgs. Retail Price: $11.99

<u>The Golden Trail: More Stories From Oregon's Mining Years</u> - In his follow-up to "Gold Dust: Stories of Oregon's Mining Years", this time around, Jackson brings us twelve tales from Oregon's Gold Rush, including the story about the first gold strike on Canyon Creek in Grant County, about the old timers who found gold by the pail full at the Victor Mine near Galice, how Iradel Bray discovered a rich ledge of gold on the Coquille River during the height of the Rogue River War, a tale of two elderly miners on the hunt for a lost mine in the Cascade Mountains, details about the discovery of the famous Armstrong Nugget and others. 5″ X 8″, 70 ppgs. Retail Price: $10.99

Oregon Mining Books

<u>Geology and Mineral Resources of Josephine County, Oregon</u> - Unavailable since the 1970's, this important publication was originally compiled by the Oregon Department of Geology and Mineral Industries and includes important details on the economic geology and mineral resources of this important mining area in South Western Oregon. Included are notes on the history, geology and development of important mines, as well as insights into the mining of gold, copper, nickel, limestone, chromium and other minerals found in large quantities in Josephine County, Oregon. 8.5″ X 11″, 54 ppgs. Retail Price: $9.99

<u>Mines and Prospects of the Mount Reuben Mining District</u> - Unavailable since 1947, this important publication was originally compiled by geologist Elton Youngberg of the Oregon Department of Geology and Mineral Industries and includes detailed descriptions, histories and the geology of the Mount Reuben Mining District in Josephine County, Oregon. Included are notes on the history, geology, development and assay statistics, as well as underground maps of all the major mines and prospects in the vicinity of this much neglected mining district. 8.5″ X 11″, 48 ppgs. Retail Price: $9.99

<u>The Granite Mining District</u> - Notes on the history, geology and development of important mines in the well known Granite Mining District which is located in Grant County, Oregon. Some of the mines discussed include the Ajax, Blue Ribbon, Buffalo, Continental, Cougar-Independence, Magnolia, New York, Standard and the Tillicum. Also included are many rare maps pertaining to the mines in the area. 8.5″ X 11″, 48 ppgs. Retail Price: $9.99

<u>Ore Deposits of the Takilma and Waldo Mining Districts of Josephine County, Oregon</u> - The Waldo and Takilma mining districts are most notable for the fact that the earliest large scale mining of placer gold and copper in Oregon took place in these two areas. Included are details about some of the earliest large gold mines in the state such as the Llano de Oro, High Gravel, Cameron, Platerica, Deep Gravel and others, as well as copper mines such as the famous Queen of Bronze mine, the Waldo, Lily and Cowboy mines. This volume also includes six maps and 20 original illustrations. 8.5″ X 11″, 74 ppgs. Retail Price: $9.99

<u>Metal Mines of Douglas, Coos and Curry Counties, Oregon</u> - Oregon mining historian Kerby Jackson introduces us to a classic work on Oregon's mining history in this important re-issue of Bulletin 14C Volume 1, otherwise known as the Douglas, Coos & Curry Counties, Oregon Metal Mines Handbook. Unavailable since 1940, this important publication was originally compiled by the Oregon Department of Geology and Mineral Industries includes detailed descriptions, histories and the geology of over 250 metallic mineral mines and prospects in this rugged area of South West Oregon. 8.5″ X 11″, 158 ppgs. Retail Price: $19.99

Metal Mines of Jackson County, Oregon - Unavailable since 1943, this important publication was originally compiled by the Oregon Department of Geology and Mineral Industries includes detailed descriptions, histories and the geology of over 450 metallic mineral mines and prospects in Jackson County, Oregon. Included are such famous gold mining areas as Gold Hill, Jacksonville, Sterling and the Upper Applegate. **8.5" X 11", 220 ppgs. Retail Price: $24.99**

Metal Mines of Josephine County, Oregon - Oregon mining historian Kerby Jackson introduces us to a classic work on Oregon's mining history in this important re-issue of Bulletin 14C, otherwise known as the Josephine County, Oregon Metal Mines Handbook. Unavailable since 1952, this important publication was originally compiled by the Oregon Department of Geology and Mineral Industries includes detailed descriptions, histories and the geology of over 500 metallic mineral mines and prospects in Josephine County, Oregon. **8.5" X 11", 250 ppgs. Retail Price: $24.99**

Metal Mines of North East Oregon - Oregon mining historian Kerby Jackson introduces us to a classic work on Oregon's mining history in this important re-issue of Bulletin 14A and 14B, otherwise known as the North East Oregon Metal Mines Handbook. Unavailable since 1941, this important publication was originally compiled by the Oregon Department of Geology and Mineral Industries and includes detailed descriptions, histories and the geology of over 750 metallic mineral mines and prospects in North Eastern Oregon. **8.5" X 11", 310 ppgs. Retail Price: $29.99**

Metal Mines of North West Oregon - Oregon mining historian Kerby Jackson introduces us to a classic work on Oregon's mining history in this important re-issue of Bulletin 14D, otherwise known as the North West Oregon Metal Mines Handbook. Unavailable since 1951, this important publication was originally compiled by the Oregon Department of Geology and Mineral Industries and includes detailed descriptions, histories and the geology of over 250 metallic mineral mines and prospects in North Western Oregon. **8.5" X 11", 182 ppgs. Retail Price: $19.99**

Mines and Prospects of Oregon - Mining historian Kerby Jackson introduces us to a classic mining work by the Oregon Bureau of Mines in this important re-issue of The Handbook of Mines and Prospects of Oregon. Unavailable since 1916, this publication includes important insights into hundreds of gold, silver, copper, coal, limestone and other mines that operated in the State of Oregon around the turn of the 19th Century. Included are not only geological details on early mines throughout Oregon, but also insights into their history, production, locations and in some cases, also included are rare maps of their underground workings. **8.5" X 11", 314 ppgs. Retail Price: $24.99**

Lode Gold of the Klamath Mountains of Northern California and South West Oregon
(See California Mining Books)

Mineral Resources of South West Oregon - Unavailable since 1914, this publication includes important insights into dozens of mines that once operated in South West Oregon, including the famous gold fields of Josephine and Jackson Counties, as well as the Coal Mines of Coos County. Included are not only geological details on early mines throughout South West Oregon, but also insights into their history, production and locations. **8.5" X 11", 154 ppgs. Retail Price: $11.99**

Chromite Mining in The Klamath Mountains of California and Oregon
(See California Mining Books)

Southern Oregon Mineral Wealth - Unavailable since 1904, this rare publication provides a unique snapshot into the mines that were operating in the area at the time. Included are not only geological details on early mines throughout South West Oregon, but also insights into their history, production and locations. Some of the mining areas include Grave Creek, Greenback, Wolf Creek, Jump Off Joe Creek, Granite Hill, Galice, Mount Reuben, Gold Hill, Galls Creek, Kane Creek, Sardine Creek, Birdseye Creek, Evans Creek, Foots Creek, Jacksonville, Ashland, the Applegate River, Waldo, Kerby and the Illinois River, Althouse and Sucker Creek, as well as insights into local copper mining and other topics. **8.5" X 11", 64 ppgs. Retail Price: $8.99**

Geology and Ore Deposits of the Takilma and Waldo Mining Districts - Unavailable since the 1933, this publication was originally compiled by the United States Geological Survey and includes details on gold and copper mining in the Takilma and Waldo Districts of Josephine County, Oregon. The Waldo and Takilma mining districts are most notable for the fact that the earliest large scale mining of placer gold and copper in Oregon took place in these two areas. Included in this report are details about some of the earliest large gold mines in the state such as the Llano de Oro, High Gravel, Cameron, Platerica, Deep Gravel and others, as well as copper mines such as the famous Queen of Bronze mine, the Waldo, Lily and Cowboy mines. In addition to geological examinations, insights are also provided into the production, day to day operations and early histories of these mines, as well as calculations of known mineral reserves in the area. This volume also includes six maps and 20 original illustrations. **8.5" X 11", 74 ppgs. Retail Price: $9.99**

Gold Mines of Oregon - Oregon mining historian Kerby Jackson introduces us to a classic work on Oregon's mining history in this important re-issue of Bulletin 61, otherwise known as "Gold and Silver In Oregon". Unavailable since 1968, this important publication was originally compiled by geologists Howard C. Brooks and Len Ramp of the Oregon Department of Geology and Mineral Industries and includes detailed descriptions, histories and the geology of over 450 gold mines Oregon. Included are notes on the history, geology and gold production statistics of all the major mining areas in Oregon including the Klamath Mountains, the Blue Mountains and the North Cascades. While gold is where you find it, as every miner knows, the path to success is to prospect for gold where it was previously found. **8.5" X 11", 344 ppgs. Retail Price: $24.99**

Mines and Mineral Resources of Curry County Oregon - Originally published in 1916, this important publication on Oregon Mining has not been available for nearly a century. Included are rare insights into the history, production and locations of dozens of gold mines in Curry County, Oregon, as well as detailed information on important Oregon mining districts in that area such as those at Agness, Bald Face Creek, Mule Creek, Boulder Creek, China Diggings, Collier Creek, Elk River, Gold Beach, Rock Creek, Sixes River and elsewhere. Particular attention is especially paid to the famous beach gold deposits of this portion of the Oregon Coast. **8.5" X 11", 140 ppgs. Retail Price: $11.99**

Chromite Mining in South West Oregon - Originally published in 1961, this important publication on Oregon Mining has not been available for nearly a century. Included are rare insights into the history, production and locations of nearly 300 chromite mines in South Western Oregon. **8.5" X 11", 184 ppgs. Retail Price: $14.99**

Mineral Resources of Douglas County Oregon - Originally published in 1972, this important publication on Oregon Mining has not been available for nearly forty years. Included are rare insights into the geology, history, production and locations of numerous gold mines and other mining properties in Douglas County, Oregon. **8.5" X 11", 124 ppgs. Retail Price: $11.99**

Mineral Resources of Coos County Oregon - Originally published in 1972, this important publication on Oregon Mining has not been available for nearly forty years. Included are rare insights into the geology, history, production and locations of numerous gold mines and other mining properties in Coos County, Oregon. **8.5" X 11", 100 ppgs. Retail Price: $11.99**

Mineral Resources of Lane County Oregon - Originally published in 1938, this important publication on Oregon Mining has not been available for nearly seventy five years. Included are extremely rare insights into the geology and mines of Lane County, Oregon, in particular in the Bohemia, Blue River, Oakridge, Black Butte and Winberry Mining Districts. **8.5" X 11", 82 ppgs. Retail Price: $9.99**

Mineral Resources of the Upper Chetco River of Oregon: Including the Kalmiopsis Wilderness - Originally published in 1975, this important publication on Oregon Mining has not been available for nearly forty years. Withdrawn under the 1872 Mining Act since 1984, real insight into the minerals resources and mines of the Upper Chetco River has long been unavailable due to the remoteness of the area. Despite this, the decades of battle between property owners and environmental extremists over the last private mining inholding in the area has continued to pique the interest of those interested in mining and other forms of natural resource use. Gold mining began in the area in the 1850's and has a rich history in this geographic area, even if the facts surrounding it are little known. Included are twenty two rare photographs, as well as insights into the Becca and Morning Mine, the Emmly Mine (also known as Emily Camp), the Frazier Mine, the Golden Dream or Higgins Mine, Hustis Mine, Peck Mine and others. **8.5" X 11", 64 ppgs. Retail Price: $8.99**

Gold Dredging in Oregon - Originally published in 1939, this important publication on Oregon Mining has not been available for nearly seventy five years. Included are extremely rare insights into the history and day to day operations of the dragline and bucketline gold dredges that once worked the placer gold fields of South West and North East Oregon in decades gone by. Also included are details into the areas that were worked by gold dredges in Josephine, Jackson, Baker and Grant counties, as well as the economic factors that impacted this mining method. This volume also offers a unique look into the values of river bottom land in relation to both farming and mining, in how farm lands were mined, re-soiled and reclamated after the dredges worked them. Featured are hard to find maps of the gold dredge fields, as well as rare photographs from a bygone era. **8.5" X 11", 86 ppgs. Retail Price: $8.99**

Quick Silver Mining in Oregon - Originally published in 1963, this important publication on Oregon Mining has not been available for over fifty years. This publication includes details into the history and production of Elemental Mercury or Quicksilver in the State of Oregon. **8.5" X 11", 238 ppgs. Retail Price: $15.99**

Mines of the Greenhorn Mining District of Grant County Oregon - Originally published in 1948, this important publication on Oregon Mining has not been available for over sixty five years. In this publication are rare insights into the mines of the famous Greenhorn Mining District of Grant County, Oregon, especially the famous Morning Mine. Also included are details on the Tempest, Tiger, Bi-Metallic, Windsor, Psyche, Big Johnny, Snow Creek, Banzette and Paramount Mines, as well as prospects in the vicinities in the famous mining areas of Mormon Basin, Vinegar Basin and Desolation Creek. Included are hard to find mine maps and dozens of rare photographs from the bygone era of Grant County's rich mining history. **8.5" X 11", 72 ppgs. Retail Price: $9.99**

Geology of the Wallowa Mountains of Oregon: Part I (Volume 1) - Originally published in 1938, this important publication on Oregon Mining has not been available for nearly seventy five years. Included are details on the geology of this unique portion of North Eastern Oregon. This is the first part of a two book series on the area. Accompanying the text are rare photographs and historic maps. 8.5" X 11", 92 ppgs. Retail Price: $9.99

Geology of the Wallowa Mountains of Oregon: Part II (Volume 2) - Originally published in 1938, this important publication on Oregon Mining has not been available for nearly seventy five years. Included are details on the geology of this unique portion of North Eastern Oregon. This is the first part of a two book series on the area. Accompanying the text are rare photographs and historic maps. 8.5" X 11", 94 ppgs. Retail Price: $9.99

Field Identification of Minerals For Oregon Prospectors - Originally published in 1940, this important publication on Oregon Mining has not been available for nearly seventy five years. Included in this volume is an easy system for testing and identifying a wide range of minerals that might be found by prospectors, geologists and rockhounds in the State of Oregon, as well as in other locales. Topics include how to put together your own field testing kit and how to conduct rudimentary tests in the field. This volume is written in a clear and concise way to make it useful even for beginners. 8.5" X 11", 158 ppgs. Retail Price: $14.99

The Bohemia Mining District of Oregon - Originally published in 1900, this important publication on Oregon Mining has not been available for over a century. Included in this volume are important insights into the famous Bohemia Mining District of Oregon, including the histories and locations of important gold mines in the area such as the Ophir Mine, Clarence, Acturas, Peek-a-boo, White Swan, Combination Mine, the Musick Mine, The California, White Ghost, The Mystery, Wall Street, Vesuvius, Story, Lizzie Bullock, Delta, Elsie Dora, Golden Slipper, Broadway, Champion Mine, Knott, Noonday, Helena, White Wings, Riverside and others. Also included are notes on the nearby Blue River Mining District. 8.5" X 11", 58 ppgs. Retail Price: $9.99

The Gold Fields of Eastern Oregon - Unavailable since 1900, this publication was originally compiled by the Baker City Chamber of Commerce Offering important insights into the gold mining history of Eastern Oregon, "The Gold Fields of Eastern Oregon" sheds a rare light on many of the gold mines that were operating at the turn of the 19th Century in Baker County and Grant County in North Eastern Oregon. Some of the areas featured include the Cable Cove District, Baisely-Elhorn, Granite, Red Boy, Bonanza, Susanville, Sparta, Virtue, Vaughn, Sumpter, Burnt River, Rye Valley and other mining districts. Included is basic information on not only many gold mines that are well known to those interested in Eastern Oregon mining history, but also many mines and prospects which have been mostly lost to the passage of time. Accompanying are numerous rare photos 8.5" X 11", 78 ppgs. Retail Price: $10.99

Gold Mining in Eastern Oregon - Originally published in 1938, this important publication on Oregon Mining has not been available for over a century. Included in this volume are important insights into the famous mining districts of Eastern Oregon during the late 1930's. Particular attention is given to those gold mines with milling and concentrating facilities in the Greenhorn, Red Boy, Alamo, Bonanza, Granite, Cable Cove, Cracker Creek, Virtue, Keating, Medical Springs, Sanger, Sparta, Chicken Creek, Mormon Basin, Connor Creek, Cornucopia and the Bull Run Mining Districts. Some of the mines featured include the Ben Harrison, North Pole-Columbia, Highland Maxwell, Baisley-Elkhorn, White Swan, Balm Creek, Twin Baby, Gem of Sparta, New Deal, Gleason, Gifford-Johnson, Cornucopia, Record, Bull Run, Orion and others. Of particular interest are the mill flow sheets and descriptions of milling operations of these mines. 8.5" X 11", 68 ppgs. Retail Price: $8.99

The Gold Belt of the Blue Mountains of Oregon - Originally published in 1901, this important publication on Oregon Mining has not been available for over a century. Included in this volume are rare insights into the gold deposits of the Blue Mountains of North East Oregon, including the history of their early discovery and early production. Extensive details are offered on this important mining area's mineralogy and economic geology, as well as insights into nearby gold placers, silver deposits and copper deposits. Featured are the Elkhorn and Rock Creek mining districts, the Pocahontas district, Auburn and Minersville districts, Sumpter and Cracker Creek, Cable Cove, the Camp Carson district, Granite, Alamo, Greenhorn, Robinsonville, the Upper Burnt River Valley and Bonanza districts, Susanville, Quartzburg, Canyon Creek, Virtue, the Copper Butte district, the North Powder River, Sparta, Eagle Creek, Cornucopia, Pine Creek, Lower Powder River, the Upper Snake River Canyon, Rye Valley, Lower Burnt River Valley, Mormon Basin, the Malheur and Clarks Creek districts, Sutton Creek and others. Of particular interest are important details on numerous gold mines and prospects in these mining districts, including their locations, histories, geology and other important information, as well as information on silver, copper and fire opal deposits. 8.5" X 11", 250 ppgs. Retail Price: $24.99

<u>Mining in the Cascades Range of Oregon</u> - Originally published in 1938, this important publication on Oregon Mining has not been available for over seventy five years. Included in this volume are rare insights into the gold mines and other types of metal mines in the Cascades Mountain Range of Oregon. Some of the important mining areas covered include the famous Bohemia Mining District, the North Santiam Mining District, Quartzville Mining District, Blue River Mining District, Fall Creek Mining District, Oakridge District, Zinc District, Buzzard-Al Sarena District, Grand Cove, Climax District and Barron Mining District. Of particular interest are important details on over 100 mines and prospects in these mining districts, including their locations, histories, geology and other important information. **8.5" X 11", 170 ppgs. Retail Price: $14.99**

<u>Beach Gold Placers of the Oregon Coast</u> - Originally published in 1934, this important publication on Oregon Mining has not been available for over 80 years. Included in this volume are rare insights into the beach gold deposits of the State of Oregon, including their locations, occurance, composition and geology. Of particular interest is information on placer platinum in Oregon's rich beach deposits. Also included are the locations and other information on some famous Oregon beach mines, including the Pioneer, Eagle, Chickamin, Iowa and beach placer mines north of the mouth of the Rogue River. **8.5" X 11", 60 ppgs. Retail Price: $8.99**

Idaho Mining Books

Gold in Idaho - Unavailable since the 1940's, this publication was originally compiled by the Idaho Bureau of Mines and includes details on gold mining in Idaho. Included is not only raw data on gold production in Idaho, but also valuable insight into where gold may be found in Idaho, as well as practical information on the gold bearing rocks and other geological features that will assist those looking for placer and lode gold in the State of Idaho. This volume also includes thirteen gold maps that greatly enhance the practical usability of the information contained in this small book detailing where to find gold in Idaho. **8.5" X 11", 72 ppgs. Retail Price: $9.99**

Geology of the Couer D'Alene Mining District of Idaho - Unavailable since 1961, this publication was originally compiled by the Idaho Bureau of Mines and Geology and includes details on the mining of gold, silver and other minerals in the famous Coeur D'Alene Mining District in Northern Idaho. Included are details on the early history of the Coeur D'Alene Mining District, local tectonic settings, ore deposit features, information on the mineral belts of the Osburn Fault, as well as detailed information on the famous Bunker Hill Mine, the Dayrock Mine, Galena Mine, Lucky Friday Mine and the infamous Sunshine Mine. This volume also includes sixteen hard to find maps. **8.5" X 11", 70 ppgs. Retail Price: $9.99**

The Gold Camps and Silver Cities of Idaho - Originally published in 1963, this important publication on Idaho Mining has not been available for nearly fifty years. Included are rare insights into the history of Idaho's Gold Rush, as well as the mad craze for silver in the Idaho Panhandle. Documented in fine detail are the early mining excitements at Boise Basin, at South Boise, in the Owyhees, at Deadwood, Long Valley, Stanley Basin and Robinson Bar, at Atlanta, on the famous Boise River, Volcano, Little Smokey, Banner, Boise Ridge, Hailey, Leesburg, Lemhi, Pearl, at South Mountain, Shoup and Ulysses, Yellow Jacket and Loon Creek. The story follows with the appearance of Chinese miners at the new mining camps on the Snake River, Black Pine, Yankee Fork, Bay Horse, Clayton, Heath, Seven Devils, Gibbonsville, Vienna and Sawtooth City. Also included are special sections on the Idaho Lead and Silver mines of the late 1800's, as well as the mining discoveries of the early 1900's that paved the way for Idaho's modern mining and mineral industry. Lavishly illustrated with rare historic photos, this volume provides a one of a kind documentary into Idaho's mining history that is sure to be enjoyed by not only modern miners and prospectors who still scour the hills in search of nature's treasures, but also those enjoy history and tromping through overgrown ghost towns and long abandoned mining camps. **8.5" X 11", 186 ppgs. Retail Price: $14.99**

Ore Deposits and Mining in North Western Custer County Idaho - Unavailable since 1913, this important publication was originally published by the Us Department of the Interior and has been unavailable for a century. Included are fine details on the geology, geography, gold placers and gold and silver bearing quartz veins of the mining region of North West Custer County, Idaho. Of particular interest is a rare look at the mines and prospects of the region, including those such as the Ramshorn Mine, SkyLark, Riverview, Excelsior, Beardsley, Pacific, Hoosier, Silver Brick, Forest Rose and dozens of others in the Bay Horse Mining District. Also covered are the mines of the Yankee Fork District such as the Lucky Boy, Badger, Black, Enterprise, Charles Dickens, Morrison, Golden Sunbeam, Montana, Golden Gate and others, as well as those in the Loon Mining District. **8.5" X 11", 126 ppgs. Retail Price: $12.99**

Gold Rush To Idaho - Unavailable since 1963, this important publication was originally published by the Idaho Bureau of Mines and has been unavailable for 50 years. "Gold Rush To Idaho" revisits the earliest years of the discovery of gold in Idaho Territory and introduces us to the conditions that the pioneer gold seekers met when they blazed a trail through the wilderness of Idaho's mountains and discovered the precious yellow metal at Oro Fino and Pierce. Subsequent rushes followed at places like Elk City, Newsome, Clearwater Station, Florence, Warrens and elsewhere. Of particular interest is a rare look at the hardships that the first miners in Idaho met with during their day to day existences and their attempts to bring law and order to their mining camps. **8.5" X 11", 88 ppgs. Retail Price: $9.99**

The Geology and Mines of Northern Idaho and North Western Montana - Unavailable since 1909, this important publication was originally published by the Us Department of the Interior and has been unavailable for a century. Included are fine details on the geology and geography of the mining regions of Northern Idaho and North Western Montana. Of particular interest is a rare look at the mines and prospects of the region, including those in the Pine Creek Mining District, Lake Pend Oreille district, Troy Mining District, Sylvanite District, Cabinet Mining District, Prospect Mining District and the Missoula Valley. Some of the mines featured include the Iron Mountain, Silver Butte, Snowshoe, Grouse Mountain Mine and others. **8.5" X 11", 142 ppgs. Retail Price: $12.99**

Mining in the Alturas Quadrangle of Blaine County Idaho - Unavailable since 1922, this important publication was originally published by the Idaho Bureau of Mines and has been unavailable for ninety years. Topics include the geology, rock formations and the formation of ore deposits in this important mining area of Idaho. Of particular focus is information on the local geology, quartz veins and ore deposits of this portion of Idaho. Included are hard to find details, including the descriptions and locations of numerous gold and silver mines in the area including the Silver King, Pilgrim, Columbia, Lone Jack, Sunbeam, Pride of the West, Lucky Boy, Scotia, Atlanta, Beaver-Bidwell and others mines and prospects. **8.5" X 11", 56 ppgs. Retail Price: $8.99**

Mining in Lemhi County Idaho - Originally published in 1913, this important book on Idaho Mining has not been available to miners for over a century. Included are rare insights into hundreds of gold, silver, copper and other mines in this famous Idaho mining area. Details include the locations, geology, history, production and other facts of the mines of this region, not only gold and silver hardrock mines, but also gold placer mines, lead-silver deposits, copper mines, cobalt-nickel deposits, tungsten and tin mines . It is lavishly illustrated with hard to find photos of the period and rare mining maps. Some of the vicinities featured include the Nicholia Mining District, Spring Mountain District, Texas District, Blue Wing District, Junction District, McDevitt District, Pratt Creek, Eldorado District, Kirtley Creek, Carmen Creek, Gibbonsville, Indian Creek, Mineral Hill District, Mackinaw, Eureka District, Blackbird District, YellowJacket District, Gravel Range District, Junction District, Parker Mountain and other mining districts. **8.5" X 11", 226 ppgs. Retail Price: $19.99**

Utah Mining Books

Fluorite in Utah - Unavailable since 1954, this publication was originally compiled by the USGS, State of Utah and U.S. Atomic Energy Commission and details the mining of fluorspar, also known as fluorite in the State of Utah. Included are details on the geology and history of fluorspar (fluorite) mining in Utah, including details on where this unique gem mineral may be found in the State of Utah. **8.5" X 11", 60 ppgs. Retail Price: $8.99**

California Mining Books

The Tertiary Gravels of the Sierra Nevada of California - Mining historian Kerby Jackson introduces us to a classic mining work by Waldemar Lindgren in this important re-issue of The Tertiary Gravels of the Sierra Nevada of California. Unavailable since 1911, this publication includes details on the gold bearing ancient river channels of the famous Sierra Nevada region of California. **8.5" X 11", 282 ppgs. Retail Price: $19.99**

The Mother Lode Mining Region of California - Unavailable since 1900, this publication includes details on the gold mines of California's famous Mother Lode gold mining area. Included are details on the geology, history and important gold mines of the region, as well as insights into historic mining methods, mine timbering, mining machinery, mining bell signals and other details on how these mines operated. Also included are insights into the gold mines of the California Mother Lode that were in operation during the first sixty years of California's mining history. **8.5" X 11", 176 ppgs. Retail Price: $14.99**

Lode Gold of the Klamath Mountains of Northern California and South West Oregon - Unavailable since 1971, this publication was originally compiled by Preston E. Hotz and includes details on the lode mining districts of Oregon and California's Klamath Mountains. Included are details on the geology, history and important lode mines of the French Gulch, Deadwood, Whiskeytown, Shasta, Redding, Muletown, South Fork, Old Diggings, Dog Creek (Delta), Bully Choop (Indian Creek), Harrison Gulch, Hayfork, Minersville, Trinity Center, Canyon Creek, East Fork, New River, Denny, Liberty (Black Bear), Cecilville, Callahan, Yreka, Fort Jones and Happy Camp mining districts in California, as well as the Ashland, Rogue River, Applegate, Illinois River, Takilma, Greenback, Galice, Silver Peak, Myrtle Creek and Mule Creek districts of South Western Oregon. Also included are insights into the mineralization and other characteristics of this important mining region. **8.5" X 11", 100 ppgs. Retail Price: $10.99**

Mines and Mineral Resources of Shasta County, Siskiyou County, Trinity County: California - Unavailable since 1915, this publication was originally compiled by the California State Mining Bureau and includes details on the gold mines of this area of Northern California. Also included are insights into the mineralization and other characteristics of this important mining region, as well as the location of historic gold mines. 8.5" X 11", 204 ppgs. Retail Price: $19.99

Geology of the Yreka Quadrangle, Siskiyou County, California - Unavailable since 1977, this publication was originally compiled by Preston E. Hotz and includes details on the geology of the Yreka Quadrangle of Siskiyou County, California. Also included are insights into the mineralization and other characteristics of this important mining region. 8.5" X 11", 78 ppgs. Retail Price: $7.99

Mines of San Diego and Imperial Counties, California - Originally published in 1914, this important publication on California Mining has not been available for a century. This publication includes important information on the early gold mines of San Diego and Imperial County, which were some of the first gold fields mined in California by early Spanish and Mexican miners before the 49ers came on the scene. Included are not only details on early mining methods in the area, production statistics and geological information, but also the location of the early gold mines that helped make California "The Golden State". Also included are details on the mining of other minerals such as silver, lead, zinc, manganese, tungsten, vanadium, asbestos, barite, borax, cement, clay, dolomite, fluospar, gem stones, graphite, marble, salines, petroleum, stronium, talc and others. 8.5" X 11", 116 ppgs. Retail Price: $12.99

Mines of Sierra County, California - Unavailable since 1920, this publication was originally compiled by the California State Mining Bureau and includes details on the gold mines of Sierra County, California. Also included are insights into the mineralization and other characteristics of this important mining region, as well as the location of historic gold mines. 8.5" X 11", 156 ppgs. Retail Price: $19.99

Mines of Plumas County, California - Unavailable since 1918, this publication was originally compiled by the California State Mining Bureau and includes details on the gold mines of Plumas County, California. Also included are insights into the mineralization and other characteristics of this important mining region, as well as the location of historic gold mines. 8.5" X 11", 200 ppgs. Retail Price: $19.99

Mines of El Dorado, Placer, Sacramento and Yuba Counties, California - Originally published in 1917, this important publication on California Mining has not been available for nearly a century. This publication includes important information on the early gold mines of El Dorado County, Placer County, Sacramento County and Yuba County, which were some of the first gold fields mined by the Forty-Niners during the California Gold Rush. Included are not only details on early mining methods in the area, production statistics and geological information, but also the location of the early gold mines that helped make California "The Golden State". Also included are insights into the early mining of chrome, copper and other minerals in this important mining area. 8.5" X 11", 204 ppgs. Retail Price: $19.99

Mines of Los Angeles, Orange and Riverside Counties, California - Originally published in 1917, this important publication on California Mining has not been available for nearly a century. This publication includes important information on the early gold mines of Los Angeles County, Orange County and Riverside County, which were some of the first gold fields mined in California by early Spanish and Mexican miners before the 49ers came on the scene. Included are not only details on early mining methods in the area, production statistics and geological information, but also the location of the early gold mines that helped make California "The Golden State". 8.5" X 11", 146 ppgs. Retail Price: $12.99

Mines of San Bernadino and Tulare Counties, California - Originally published in 1917, this important publication on California Mining has not been available for nearly a century. This publication includes important information on the early gold mines of San Bernadino and Tulare County, which were some of the first gold fields mined in California by early Spanish and Mexican miners before the 49ers came on the scene. Included are not only details on early mining methods in the area, production statistics and geological information, but also the location of the early gold mines that helped make California "The Golden State". Also included are details on the mining of other minerals such as copper, iron, lead, zinc, manganese, tungsten, vanadium, asbestos, barite, borax, cement, clay, dolomite, fluospar, gem stones, graphite, marble, salines, petroleum, stronium, talc and others. 8.5" X 11", 200 ppgs. Retail Price: $19.99

Chromite Mining in The Klamath Mountains of California and Oregon - Unavailable since 1919, this publication was originally compiled by J.S. Diller of the United States Department of Geological Survey and includes details on the chromite mines of this area of Northern California and Southern Oregon. Also included are insights into the mineralization and other characteristics of this important mining region, as well as the location of historic mines. Also included are insights into chromite mining in Eastern Oregon and Montana. 8.5" X 11", 98 ppgs. Retail Price: $9.99

Mines and Mining in Amador, Calaveras and Tuolumne Counties, California - Unavailable since 1915, this publication was originally compiled by William Tucker and includes details on the mines and mineral resources of this important California mining area. Included are details on the geology, history and important gold mines of the region, as well as insights into other local mineral resources such as asbestos, clay, copper, talc, limestone and others. Also included are insights into the mineralization and other characteristics of this important portion of California's Mother Lode mining region. 8.5" X 11", 198 ppgs. **Retail Price: $14.99**

The Cerro Gordo Mining District of Inyo County California - Unavailable since 1963, this publication was originally compiled by the United States Department of Interior. Included are insights into the mineralization and other characteristics of this important mining region of Southern California. Topics include the mining of gold and silver in this important mining district in Inyo County, California, including details on the history, production and locations of the Cerro Gordo Mine, the Morning Star Mine, Estelle Tunnel, Charles Lease Tunnel, Ignacio, Hart, Crosscut Tunnel, Sunset, Upper Newtown, Newtown, Ella, Perseverance, Newsboy, Belmont and other silver and gold mines in the Cerro Gordo Mining District. This volume also includes important insights into the fossil record, geologic formations, faults and other aspects of economic geology in this California mining district. 8.5" X 11", 104 ppgs. **Retail Price: $10.99**

Mining in Butte, Lassen, Modoc, Sutter and Tehama Counties of California - Unavailable since 1917, this publication was originally compiled by the United States Department of Interior. Included are insights into the mineralization and other characteristics of this important mining region of California. Topics include the mining of asbestos, chromite, gold, diamonds and manganese in Butte County, the mining of gold and copper in the Hayden Hill and Diamond Mountain mining districts of Lassen County, the mining of coal, salt, copper and gold in the High Grade and Winters mining districts of Modoc County, gold mining in Sutter County and the mining of gold, chromite, manganese and copper in Tehama County. This volume also includes the production records and locations of numerous mines in this important mining region. 8.5" X 11", 114 ppgs. **Retail Price: $11.99**

Mines of Trinity County California - Originally published in 1965, this important publication on California Mining has not been available for nearly fifty years. This publication includes important information on mines and mining in Trinity County, California, as well insights into the mineralization and geology of this important mining area in Northern California. Included are extensive details on hardrock and placer gold mines and prospects, including charts showing the locations of these historic mines.. 8.5" X 11", 144 ppgs. **Retail Price: $12.99**

Mines of Kern County California - Originally published in 1962, this important publication on California Mining has not been available for nearly fifty years. This publication includes important information on mines and mining in Kern County, California, as well insights into the mineralization and geology of this important mining area in California. Included are extensive details on hardrock and placer gold mines and prospects, including charts showing the locations of these historic mines. 8.5" X 11", 398 ppgs. **Retail Price: $24.99**

Mines of Calaveras County California - Originally published in 1962, this important publication on California Mining has not been available for nearly fifty years. This publication includes important information on mines and mining in Calaveras County, California, as well insights into the mineralization and geology of this important mining area in Northern California. Included are extensive details on hardrock and placer gold mines and prospects, including charts showing the locations of these historic mines. 8.5" X 11", 236 ppgs. **Retail Price: $19.99**

Lode Gold Mining in Grass Valley California - Unavailable since 1940, this publication was originally compiled by the United States Department of Interior. Included are insights into the gold mineralization and other characteristics of this important mining region of Nevada County, California. This volume also includes important insights into the geologic formations, faults and other aspects of economic geology in this California mining district. Of particular interest are the fine details on many hardrock gold mines in the area, including their locations, histories, development and mineralization. Some of the mines featured include the Gold Hill Mine, Massachusetts Hill, Boundary, Peabody, Golden Center, North Star, Omaha, Lone Jack, Homeward Bound, Hartery, Wisconsin, Allison Ranch, Phoenix, Kate Hayes, W.Y.O.D., Empire, Rich Hill, Daisy Hill, Orleans, Sultana, Centennial, Conlin, Ben Franklin, Crown Point and many others. 8.5" X 11", 148 ppgs. **Retail Price: $12.99**

Lode Mining in the Alleghany District of Sierra County California - Unavailable since 1913, this publication was originally compiled by the United States Department of Interior. Included are insights into the mineralization and other characteristics of this important mining region of Sierra County. Included are details on the history, production and locations of numerous hardrock gold mines in this famous California area, including the Tightner Mine, Minnie D., Osceola, Eldorado, Twenty One, Sherman, Kenton, Oriental, Rainbow, Plumbago, Irelan, Gold Canyon, North Fork, Federal, Kate Hardy and others. This volume also includes important insights into the fossil record, geologic formations, faults and other aspects of economic geology in this California mining district. 8.5" X 11", 48 ppgs. **Retail Price: $7.99**

Six Months In The Gold Mines During The California Gold Rush - Unavailable since 1850, this important work is a first hand account of one "49'ers" personal experience during the great California Gold Rush, shedding important light on one of the most exciting periods in the history of not only California, but also the world. Compiled from journals written between 1847 and 1849 by E. Gould Buffum, a native of New York, "Six Months In The Gold Mines During The California Gold Rush" offers a rare look into the day to day lives of the people who came to California to work in her gold mines when the state was still a great frontier. **8.5" X 11", 290 ppgs. Retail Price: $19.99**

Quartz Mines of the Grass Valley Mining District of California - Unavailable since 1867, this important publication has not been available since those days. This rare publication offers a short dissertation on the early hardrock mines in this important mining district in the California Mother Lode region between the 1850's and 1860's. Also included are hard to find details on the mineralization and locations of these mines, as well as how they were operated in those day. **8.5" X 11", 44 ppgs. Retail Price: $8.99**

Alaska Mining Books

Ore Deposits of the Willow Creek Mining District, Alaska - Unavailable since 1954, this hard to find publication includes valuable insights into the Willow Creek Mining District near Hatcher Pass in Alaska. The publication includes insights into the history, geology and locations of the well known mines in the area, including the Gold Cord, Independence, Fern, Mabel, Lonesome, Snowbird, Schroff-O'Neil, High Grade, Marion Twin, Thorpe, Webfoot, Kelly-Willow, Lane, Holland and others. **8.5" X 11", 96 ppgs. Retail Price: $9.99**

The Juneau Gold Belt of Alaska - Unavailable since 1906, this hard to find publication includes valuable insights into the gold mines around Juneau, Alaska. The publication includes important details into the history, geology and locations of the well known gold mines and prospects in the area, including those around Windham Bay, Holkham Bay, Port Snettisham, on Grindstone and Rhine Creeks, Gold Creek, Douglas Island, Salmon Creek, Lemon Creek, Nugget Creek, from the Mendenhall River to Berners Bay, McGinnis Creek, Montana Creek, Peterson Creek, Windfall Creek, the Eagle River, Yankee Basin, Yankee Curve, Kowee Creek and elsewhere. Not only are gold placer mines included, but also hardrock gold mines. **8.5" X 11", 224 ppgs. Retail Price: $19.99**

Arizona Mining Books

Mines and Mining in Northern Yuma County Arizona - Originally published in 1911, this important publication on Arizona Mining has not been available for over a hundred years. Included are rare insights into the gold, silver, copper and quicksilver mines of Yuma County, Arizona together with hard to find maps and photographs. Some of the mines and mining districts featured include the Planet Copper Mine, Mineral Hill, the Clara Consolidated Mine, Viati Mine, Copper Basin prospect, Bowman Mine, Quartz King, Billy Mack, Carnation, the Wardwell and Osbourne, Valensuella Copper, the Mariquita, Colonial Mine, the French American, the New York-Plomosa, Guadalupe, Lead Camp, Mudersbach Copper Camp, Yellow Bird, the Arizona Northern (Salome Strike), Bonanza (Harqua Hala), Golden Eagle, Hercules, Socorro and others. **8.5" X 11", 144 ppgs. Retail Price: $11.99**

The Aravaipa and Stanley Mining Districts of Graham County Arizona - Originally published in 1925, this important publication on Arizona Mining has not been available for nearly ninety years. Included are rare insights into the gold and silver mines of these two important mining districts, together with hard to find maps. **8.5" X 11", 140 ppgs. Retail Price: $11.99**

Gold in the Gold Basin and Lost Basin Mining Districts of Mohave County, Arizona - This volume contains rare insights into the geology and gold mineralization of the Gold Basin and Lost Basin Mining Districts of Mohave County, Arizona that will be of benefit to miners and prospectors. Also included is a significant body of information on the gold mines and prospects of this portion of Arizona. This volume is lavishly illustrated with rare photos and mining maps. **8.5" X 11", 188 ppgs. Retail Price: $19.99**

Mines of the Jerome and Bradshaw Mountains of Arizona - This important publication on Arizona Mining has not been available for ninety years. This volume contains rare insights into the geology and ore deposits of the Jerome and Bradshaw Mountains of Arizona that will be of benefit to miners and prospectors who work those areas. Included is a significant body of information on the mines and prospects of the Verde, Black Hills, Cherry Creek, Prescott, Walker, Groom Creek, Hassayampa, Bigbug, Turkey Creek, Agua Fria, Black Canyon, Peck, Tiger, Pine Grove, Bradshaw, Tintop, Humbug and Castle Creek Mining Districts. This volume is lavishly illustrated with rare photos and mining maps. **8.5" X 11", 218 ppgs. Retail Price: $19.99**

The Ajo Mining District of Pima County Arizona - This important publication on Arizona Mining has not been available for nearly seventy years. This volume contains rare insights into the geology and mineralization of the Ajo Mining District in Pima County, Arizona and in particular the famous New Cornelia Mine. **8.5" X 11", 126 ppgs. Retail Price: $11.99**

Mining in the Santa Rita and Patagonia Mountains of Arizona - Originally published in 1915, this important publication on Arizona Mining has not been available for nearly a century. Included are rare insights into hundreds of gold, silver, copper and other mines in this famous Arizona mining area. Details include the locations, geology, history, production and other facts of the mines of this region. 8.5" X 11", 394 ppgs. Retail Price: $24.99

Mining in the Bisbee Quadrangle of Arizona - Originally published in 1906, this important publication on Arizona Mining has not been available for nearly a century. Included are rare insights into hundreds of gold, silver, copper and other mines in this famous Arizona mining area. Details include the locations, geology, history, production and other facts of the mines of this important mining region. 8.5" X 11", 188 ppgs. Retail Price: $14.99

Montana Mining Books

A History of Butte Montana: The World's Greatest Mining Camp - First published in 1900 by H.C. Freeman, this important publication sheds a bright light on one of the most important mining areas in the history of The West. Together with his insights, as well as rare photographs of the periods, Harry Freeman describes Butte and its vicinity from its early beginnings, right up to its flush years when copper flowed from its mines like a river. At the time of publication, Butte, Montana was known worldwide as "The Richest Mining Spot On Earth" and produced not only vast amounts of copper, but also silver, gold and other metals from its mines. Freeman illustrates, with great detail, the most important mines in the vicinity of Butte, providing rare details on their owners, their history and most importantly, how the mines operated and how their treasures were extracted. Of particular interest are the dozens of rare photographs that depict mines such as the famous Anaconda, the Silver Bow, the Smoke House, Moose, Paulin, Buffalo, Little Minah, the Mountain Consolidated, West Greyrock, Cora, the Green Mountain, Diamond, Bell, Parnell, the Neversweat, Nipper, Original and many others. 8.5" X 11", 142 ppgs. Retail Price: $12.99

The Butte Mining District of Montana - This important publication on Montana Mining has not been available for over a century. Included are rare insights into the gold, copper and silver mines of Butte, Montana together with hard to find maps and photographs. Some of the topics include the early history of gold, silver and copper mining in the Butte area, insight into the geology of its mining areas, the local distribution of gold, silver and copper ores, as well their composition and how to identify them. Also included are detailed facts about the mines in the Butte Mining District, including the famous Anaconda Mine, Gagnon, Parrot, Blue Vein, Moscow, Poulin, Stella, Buffalo, Green Mountain, Wake Up Jim, the Diamond-Bell Group, Mountain Consolidated, East Greyrock, West Greyrock, Snowball, Corra, Speculator, Adirondack, Miners Union, the Jessie-Edith May Group, Otisco, Iduna, Colorado, Lizzie, Cambers, Anderson, Hesperus, Preferencia and dozens of others. 8.5" X 11", 298 ppgs. Retail Price: $24.99

Mines of the Helena Mining Region of Montana - This important publication on Montana Mining has not been available for over a century. Included are rare insights into the gold, copper and silver mines of the vicinity of Helena, Montana, including the Marysville Mining District, Elliston Mining District, Rimini Mining District, Helena Mining District, Clancy Mining District, Wickes Mining District, Boulder and Basin Mining Districts and the Elkhorn Mining District. Some of the topics include the early history of gold, silver and copper mining in the Helena area, insight into the geology of its mining areas, the local distribution of gold, silver and copper ores, as well their composition and how to identify them. Also included are detailed facts, history, geology and locations of over one hundred gold, silver and copper mines in the area . 8.5" X 11", 162 ppgs, Retail Price: $14.99

Mines and Geology of the Garnet Range of Montana - This important publication on Montana Mining has not been available for over a century. Included are rare insights into the gold, copper and silver mines of the vicinity of this important mining area of Montana. Some of the topics include the early history of gold, silver and copper mining in the Garnet Mountains, insight into the geology of its mining areas, the local distribution of gold, silver and copper ores, as well their composition and how to identify them. Also included are detailed facts, history, geology and locations of numerous gold, silver and copper mines in the area . 8.5" X 11", 100 ppgs, Retail Price: $11.99

Mines and Geology of the Philipsburg Quadrangle of Montana - This important publication on Montana Mining has not been available for over a century. Included are rare insights into the gold, copper and silver mines of the vicinity of this important mining area of Montana. Some of the topics include the early history of gold, silver and copper mining in the Philipsburg Quadrangle, insight into the geology of its mining areas, the local distribution of gold, silver and copper ores, as well their composition and how to identify them. Also included are detailed facts, history, geology and locations of over one hundred gold, silver and copper mines in the area 8.5" X 11", 290 ppgs, Retail Price: $24.99

Geology of the Marysville Mining District of Montana - Included are rare insights into the mining geology of the Marysville Mining District. Some of the topics include the early history of gold, silver and copper mining in the area, insight into the geology of its mining areas, the local distribution of gold, silver and copper ores, as well their composition and how to identify them. Also included are detailed facts, history, geology and locations of gold, silver and copper mines in the area 8.5" X 11", 198 ppgs, Retail Price: $19.99

The Geology and Mines of Northern Idaho and North Western Montana

See listing under Idaho.

Nevada Mining Books

The Bull Frog Mining District of Nevada - Unavailable since 1910, this publication was originally compiled by the United States Department of Interior. This volume also includes important insights into the geologic formations, faults and other aspects of economic geology in this Nevada mining district. Of particular interest are the fine details on many mines in the area, including their locations, histories, development and mineralization. Some of the mines featured include the National Bank Mine, Providence, Gibraltor, Tramps, Denver, Original Bullfrog, Gold Bar, Mayflower, Homestake-King and other mines and prospects. **8.5″ X 11″, 152 ppgs, Retail Price: $14.99**

History of the Comstock Lode - Unavailable since 1876, this publication was originally released by John Wiley & Sons. This volume also includes important insights into the famous Comstock Lode of Nevada that represented the first major silver discovery in the United States. During its spectacular run, the Comstock produced over 192 million ounces of silver and 8.2 million ounces of gold. Not only did the Comstock result in one of the largest mining rushes in history and yield immense fortunes for its owners, but it made important contributions to the development of the State of Nevada, as well as neighboring California. Included here are important details on not only the early development and history of the Comstock, but also rare early insight into its mines, ore and its geology. **8.5″ X 11″, 244 ppgs, Retail Price: $19.99**

Colorado Mining Books

Ores of The Leadville Mining District - Unavailable since 1926, this publication was originally compiled by the United States Department of Interior. This volume also includes important insights into the ores and mineralization of the Leadville Mining District in Colorado. Topics include historic ore prospecting methods, local geology, insights into ore veins and stockworks, the local trend and distribution of ore channels, reverse faults, shattered rock above replacement ore bodies, mineral enrichment in oxidized and sulphide zones and more. **8.5″ X 11″, 66 ppgs, Retail Price: $8.99**

Mining in Colorado - Unavailable since 1926, this publication was originally compiled by the United States Department of Interior. This volume also includes important insights into the mining history of Colorado from its early beginnings in the 1850's right up to the mid 1920's. Not only is Colorado's gold mining heritage included, but also its silver, copper, lead and zinc mining industry. Each mining area is treated separately, detailing the development of Colorado's mines on a county by county basis. **8.5″ X 11″, 284 ppgs, Retail Price: $19.99**

Gold Mining in Gilpin County Colorado - Unavailable since 1876, this publication was originally compiled by the Register Steam Printing House of Central City, Colorado. A rare glimpse at the gold mining history and early mines of Gilpin County, Colorado from their first discovery in the 1850's up to the "flush years" of the mid 1870's. Of particular interest is the history of the discovery of gold in Gilpin County and details about the men who made those first strikes. Special focus is given to the early gold mines and first mining districts of the area, many of which are not detailed in other books on Colorado's gold mining history. **8.5″ X 11″, 156 ppgs, Retail Price: $12.99**

Mining in the Gold Brick Mining District of Colorado - Important insights into the history of the Gold Brick Mining District, as well as its local geography and economic geology. Also included are the histories and locations of historic mines in this important Colorado Mining District, including the Cortland, Carter, Raymond, Gold Links, Sacramento, Bassick, Sandy Hook, Chronicle, Grand Prize, Chloride, Granite Mountain, Lucille, Gray Mountain, Hilltop, Maggie Mitchell, Silver Islet, Revenue, Roosevelt, Carbonate King and others. In addition to hardrock mining, are also included are details on gold placer mining in this portion of Colorado. **8.5″ X 11″, 140 ppgs, Retail Price: $12.99**

Washington Mining Books

The Republic Mining District of Washington - Unavailable since 1910, this important publication was originally published by the Washington Geologic Survey and has been unavailable for a century. Topics include the geology, rock formations and the formation of ore deposits in this important mining area of Washington State. Also included are hard to find details on the geology, history and locations of dozens of mines in the area. Some of the mines featured include the New Republic Mine, Ben Hur, Morning Glory, the South Republic Mine, Quilp, Surprise, Black Tail, Lone Pine, San Poil, Mountain Lion, Tom Thumb, Elcaliph and many others. **8.5″ X 11″, 94 ppgs, Retail Price: $10.99**

The Myers Creek and Nighthawk Mining Districts of Washington - Unavailable since 1911, this important publication was originally published by the Washington Geologic Survey and has been unavailable for a century. Topics include the geology, rock formations and the formation of ore deposits in these important mining areas of Washington State. Also included are hard to find details on the geology, history and locations of dozens of mines in the area. Some of the mines featured include the Grant Mine, Monterey, Nip and Tuck, Myers Creek, Number Nine, Neutral, Rainbow, Aztec, Crystal Butte, Apex, Butcher Boy, Molson, Mad River, Olentangy, Delate, Kelsey, Golden Chariot, Okanogan, Ohio, Forty-Ninth Parallel, Nighthawk, Favorite, Little Chopaka, Summit, Number One, California, Peerless, Caaba, Prize Group, Ruby, Mountain Sheep, Golden Zone, Rich Bar, Similkameen, Kimberly, Triune, Hiawatha, Trinity, Hornsilver, Maquae, Bellevue, Bullfrog, Palmer Lake, Ivanhoe, Copper World and many others.
   8.5" X 11", 136 ppgs, Retail Price: $12.99

The Blewett Mining District of Washington - Unavailable since 1911, this important publication was originally published by the Washington Geologic Survey and has been unavailable for a century. Topics include the geology, rock formations and the formation of ore deposits in this important mining area of Washington State. Also included are hard to find details on the geology, history and locations of dozens of mines in the area. Some of the mines featured include the Washington Meteor, Alta Vista, Pole Pick, Blinn, North Star, Golden Eagle, Tip Top, Wilder, Golden Guinea, Lucky Queen, Blue Bell, Prospect, Homestake, Lone Rock, Johnson, and others. **8.5" X 11", 134 ppgs, Retail Price: $12.99**

Silver Mining In Washington - Unavailable since 1955, this important publication was originally published by the Washington Geologic Survey. Featured are the hard to find locations and details pertaining to Washington's silver mines. **8.5" X 11", 180 ppgs, Retail Price: $15.99**

The Mines of Snohomish County Washington - Unavailable since 1942, this important publication was originally published by the Washington Geologic Survey and has been unavailable for seventy years. Featured are details on a large number of gold, silver, copper, lead and other metallic mineral mines. Included are the locations of each historic mine, along with information on the commodity produced. **8.5" X 11", 98 ppgs, Retail Price: $10.99**

The Mines of Chelan County Washington - Unavailable since 1943, this important publication was originally published by the Washington Geologic Survey and has been unavailable for seventy years. Featured are details on a large number of gold, silver, copper, lead and other metallic mineral mines. Included are the locations of each historic mine, along with information on the commodity.   **8.5" X 11", 88 ppgs, Retail Price: $9.99**

Metal Mines of Washington - Unavailable since 1921, this important publication was originally published by the Washington Geologic Survey and has been unavailable for nearly ninety years. Widely considered a masterpiece on the Washington Mining Industry, "Metal Mines of Washington" sheds light on the important details of Washington's early mining years. Featured are details on hundreds of gold, silver, copper, lead and other metallic mineral mines. Included are hard to find details on the mineral resources of this state, as well as the locations of historic mines. Lavishly illustrated with maps and historic photos and complete with a glossary to explain any technical terms found in the text, this is one of the most important works on mining in the State of Washington. No prospector or miner should be without it if they are interested in mining in Washington.   **8.5" X 11", 396 ppgs, Retail Price: $24.99**

Gem Stones In Washington - Unavailable since 1949, this important publication was originally published by the Washington Geologic Survey and has been unavailable since first published. Included are details on where to find naturally occurring gem stones in the State of Washington, including quartz crystal, amethyst, smoky quartz, milky quartz, agates, bloodstone, carnelian, chert, flint, jasper, onyx, petrified wood, opal, fire opal, hyalite and others. **8.5" X 11", 54 ppgs, Retail Price: $8.99**

The Covada Mining District of Washington - Unavailable since 1913, this important publication was originally published by the Washington Geologic Survey and has been unavailable for a century. Topics include the geology, rock formations and the formation of ore deposits in this important mining area of Washington State. Also included are hard to find details on the geology, history and locations of dozens of mines in the area. Some of the mines featured include the Admiral, Advance, Algonkian, Big Bug, Big Chief, Big Joker, Black Hawk, Black Tail, Black Thorn, Captain, Cherokee Strip, Colorado, Dan Patch, Dead Shot, Etta, Good Ore, Greasy Run, Great Scott, Idora, IXL, Jay Bird, Kentucky Bell, King Solomon, Laurel, Laura S, Little Jay, Meteor, Neglected, Northern Light, Old Nell, Plymouth Rock, Polaris, Quandary, Reserve, Shoo Fly, Silver Plume, Three Pines, Vernie, White Rose and dozens of others.   **8.5" X 11", 114 ppgs, Retail Price: $10.99**

The Index Mining District of Washington - Unavailable since 1912, this important publication was originally published by the Washington Geologic Survey and has been unavailable for a century. Topics include the geology, rock formations and the formation of ore deposits in this important mining area of Washington State. Also included are hard to find details on the geology, history and locations of dozens of mines in the area. Some of the mines featured include the Sunset, Non-Pareil, Ethel Consolidated, Kittaning, Merchant, Homestead, Co-operative, Lost Creek, Uncle Sam, Calumet, Florence-Rae, Bitter Creek, Index Peacock, Gunn Peak, Helena, North Star, Buckeye. Copper Bell, Red Cross and others. **8.5" X 11", 114 ppgs, Retail Price: $11.99**

Mining & Mineral Resources of Stevens County Washington - Unavailable since 1920, this important publication was originally published by the Washington Geologic Survey and has been unavailable for a century. Topics include the geology, rock formations and the formation of ore deposits in these important mining areas of Washington State. Also included are hard to find details on the geology, history and locations of hundreds of mines in the area. **8.5" X 11", 372 ppgs, Retail Price: $24.99**

The Mines and Geology of the Loomis Quadrangle Okanogan County, Washington - Unavailable since 1972, this important publication was originally published by the Washington Geologic Survey and has been unavailable for a century. Topics include the geology, rock formations and the formation of ore deposits in this important mining area of Washington State. Also included are hard to find details on the geology, history and locations of dozens of gold, copper, silver and other mines in the area. **8.5" X 11", 150 ppgs, Retail Price: $12.99**

The Conconully Mining District of Okanogan County Washington - Unavailable since 1973, this important publication was originally published by the Washington Geologic Survey and has been unavailable for a century. Topics include the geology, rock formations and the formation of ore deposits in this important mining area of Washington State, which also includes Salmon Creek, Blue Lake and Galena. Also included are hard to find details on the geology, mining history and locations of dozens of mines in the area. Some of the mines include Arlington, Fourth of July, Sonny Boy, First Thought, Last Chance, War Eagle-Peacock, Wheeler, Mohawk, Lone Star, Woo Loo Moo Loo, Keystone, Hughes, Plant-Callahan, Johnny Boy, Leuena, Gubser, John Arthur, Tough Nut, Homestake, Key and many others **8.5" X 11", 68 ppgs, Retail Price: $8.99**

Wyoming Mining Books

Mining in the Laramie Basin of Wyoming - Unavailable since 1909, this publication was originally compiled by the United States Department of Interior. Also included are insights into the mineralization and other characteristics of this important mining region, especially in regards to coal, limestone, gypsum, bentonite clay, cement, sand, clay and copper. **8.5" X 11", 104 ppgs, Retail Price: $11.99**

New Mexico Mining Books

The Mogollon Mining District of New Mexico - Unavailable since 1927, this important publication was originally published by the US Department of Interior and has been unavailable for 80 years. Topics include the geology, rock formations and the formation of ore deposits in this important mining area in New Mexico. Of particular focus is information on the history and production of the ore deposits in this area, their form and structure, vein filling, their paragenesis, origins and ore shoots, as well as oxidation and supergene enrichment. Also included are hard to find details, including the descriptions and locations of numerous gold, silver and other types of mines, including the Eureka, Pacific, South Alpine, Great Western, Enterprise, Buffalo, Mountain View, Floride, Gold Dust, Last Chance, Deadwood, Confidence, Maud S., Deep Down, Little Fanney, Trilby, Johnson, Alberta, Comet, Golden Eagle, Cooney, Queen, the Iron Crown, Eberle, Clifton, Andrew Jackson mine, Mascot and others. **8.5" X 11", 144 ppgs, Retail Price: $12.99**

The Percha Mining District of Kingston New Mexico - Unavailable since 1883, this important publication was originally published by the Kingston Tribune and has been unavailable for over one hundred and thirty five years. Having been written during the earliest years of gold and silver mining in the Percha Mining District, unlike other books on the subject, this work offers the unique perspective of having actually been written while the early mining history of this area was still being made. In fact, the work was written so early in the development of this area that many of the notable mines in the Percha District were less than a few years old and were still being operated by their original discoverers with the same enthusiasm as when they were first located. Included are hard to find details on the very earliest gold and silver mines of this important mining district near Kingston in Sierra County, New Mexico. **8.5" X 11", 68 ppgs, Retail Price: $9.99**

East Coast Mining Books

The Gold Fields of the Southern Appalachians - Unavailable since 1895, this important publication was originally published by the US Department of Interior and has been unavailable for nearly 120 years. Topics include the geology, rock formations and the formation of ore deposits in this important mining area of the American South. Of particular focus is information on the history and statistics of the ore deposits in this area, their form and structure and veins. Also included are details on the placer gold deposits of the region. The gold fields of the Georgian Belt, Carolinian Belt and the South Mountain Mining District of North Carolina are all treated in descriptive detail. Included are hard to find details, including the descriptions and locations of numerous gold mines in Georgia, North Carolina and elsewhere in the American South. Also included are details on the gold belts of the British Maritime Provinces and the Green Mountains. **8.5" X 11", 104 ppgs, Retail Price: $9.99**

Gold Rush Tales Series

Millions in Siskiyou County Gold - In this first volume of the "Gold Rush Tales" series, leading mining historian and editor Kerby Jackson, introduces us to the story of how millions of dollars worth of gold was discovered in Siskiyou County during the California Gold Rush. Lavishly illustrated with photos from the 19th Century, this hard to find information was first published in 1897 and sheds important light onto the gold rush era in Siskiyou County, California and the experiences of the men who dug for the gold and actually found it. **8.5" X 11", 82 ppgs, Retail Price: $9.99**

The California Rand in the Days of '49 - In this second volume of the "Gold Rush Tales" series, leading mining historian and editor Kerby Jackson, introduces us to four tales from the California Gold Rush. Lavishly illustrated with photos from the 19th Century, this hard to find information was first published in 1890's and includes the stories of "California's Rand", details about Chinese miners, how one early miner named Baker struck it rich and also the story of Alphonzo Bowers, who invented the first hydraulic gold dredge. **8.5" X 11", 54 ppgs, Retail Price: $9.99**

More Mining Books

Prospecting and Developing A Small Mine - Topics covered include the classification of varying ores, how to take a proper ore sample, the proper reduction of ore samples, alluvial sampling, how to understand geology as it is applied to prospecting and mining, prospecting procedures, methods of ore treatment, the application of drilling and blasting in a small mine and other topics that the small scale miner will find of benefit. **8.5" X 11", 112 ppgs, Retail Price: $11.99**

Timbering For Small Underground Mines - Topics covered include the selection of caps and posts, the treatment of mine timbers, how to install mine timbers, repairing damaged timbers, use of drift supports, headboards, squeeze sets, ore chute construction, mine cribbing, square set timbering methods, the use of steel and concrete sets and other topics that the small underground miner will find of benefit. This volume also includes twenty eight illustrations depicting the proper construction of mine timbering and support systems that greatly enhance the practical usability of the information contained in this small book. **8.5" X 11", 88 ppgs. Retail Price: $10.99**

Timbering and Mining - A classic mining publication on Hard Rock Mining by W.H. Storms. Unavailable since 1909, this rare publication provides an in depth look at American methods of underground mine timbering and mining methods. Topics include the selection and preservation of mine timbers, drifting and drift sets, driving in running ground, structural steel in mine workings, timbering drifts in gravel mines, timbering methods for driving shafts, positioning drill holes in shafts, timbering stations at shafts, drainage, mining large ore bodies by means of open cuts or by the "Glory Hole" system, stoping out ore in flat or low lying veins, use of the "Caving System", stoping in swelling ground, how to stope out large ore bodies, Square Set timbering on the Comstock and its modifications by California miners, the construction of ore chutes, stoping ore bodies by use of the "Block System", how to work dangerous ground, information on the "Delprat System" of stoping without mine timbers, construction and use of headframes and much more. This volume provides a reference into not only practical methods of mining and timbering that may be employed in narrow vein mining by small miners today, but also rare insights into how mines were being worked at the turn of the 19th Century. **8.5" X 11", 288 ppgs. Retail Price: $24.99**

A Study of Ore Deposits For The Practical Miner - Mining historian Kerby Jackson introduces us to a classic mining publication on ore deposits by J.P. Wallace. First published in 1908, it has been unavailable for over a century. Included are important insights into the properties of minerals and their identification, on the occurrence and origin of gold, on gold alloys, insights into gold bearing sulfides such as pyrites and arsenopyrites, on gold bearing vanadium, gold and silver tellurides, lead and mercury tellurides, on silver ores, platinum and iridium, mercury ores, copper ores, lead ores, zinc ores, iron ores, chromium ores, manganese ores, nickel ores, tin ores, tungsten ores and others. Also included are facts regarding rock forming minerals, their composition and occurrences, on igneous, sedimentary, metamorphic and intrusive rocks, as well as how they are geologically disturbed by dikes, flows and faults, as well as the effects of these geologic actions and why they are important to the miner. Written specifically with the common miner and prospector in mind, the book will help to unlock the earth's hidden wealth for you and is written in a simple and concise language that anyone can understand. **8.5" X 11", 366 ppgs. Retail Price: $24.99**

Mine Drainage - Unavailable since 1896, this rare publication provides an in depth look at American methods of underground mine drainage and mining pump systems. This volume provides a reference into not only practical methods of mining drainage that may be employed in narrow vein mining by small miners today, but also rare insights into how mines were being worked at the turn of the 19th Century. **8.5" X 11", 218 ppgs. Retail Price: $24.99**

Fire Assaying Gold, Silver and Lead Ores - Unavailable since 1907, this important publication was originally published by the Mining and Scientific Press and was designed to introduce miners and prospectors of gold, silver and lead to the art of fire assaying. Topics include the fire assaying of ores and products containing gold, silver and lead; the sampling and preparation of ore for an assay; care of the assay office, assay furnaces; crucibles and scorifiers; assay balances; metallic ores; scorification assays; cupelling; parting' crucible assays, the roasting of ores and more. This classic provides a time honored method of assaying put forward in a clear, concise and easy to understand language that will make it a benefit to even beginners. **8.5" X 11", 96 ppgs. Retail Price: $11.99**

Methods of Mine Timbering - Originally published in 1896, this important publication on mining engineering has not been available for nearly a century. Included are rare insights into historical methods of timbering structural support that were used in underground metal mines during the California that still have a practical application for the small scale hardrock miner of today. **8.5" X 11", 94 ppgs. Retail Price: $10.99**

The Enrichment of Copper Sulfide Ores - First published in 1913, it has been unavailable for over a century. Topics include the definition and types of ore enrichment, the oxidation of copper ores, the precipitation of metallic sulfides. Also included are the results of dozens of lab experiments pertaining to the enrichment of sulfide ores that will be of interest to the practical hard rock mine operator in his efforts to release the metallic bounty from his mine's ore. **8.5" X 11", 92 ppgs. Retail Price: $9.99**

A Study of Magmatic Sulfide Ores - Unavailable since 1914, this rare publication provides an in depth look at magmatic sulfide ores. Some of the topics included are the definition and classification of magmatic ores, descriptions of some magmatic sulfide ore deposits known at the time of publication including copper and nickel bearing pyrrhitic ore bodies, chalcopyrite-bornite deposits, pyritic deposits, magnetite-ileminite deposits, chromite deposits and magmatic iron ore deposits. Also included are details on how to recognize these types of ore deposits while prospecting for valuable hardrock minerals. **8.5" X 11", 138 ppgs. Retail Price: $11.99**

The Cyanide Process of Gold Recovery - Unavailable since 1894 and released under the name "The Cyanide Process: Its Practical Application and Economical Results", this rare publication provides an in depth look at the early use of cyanide leaching for gold recovery from hardrock mine ores. This volume provides a reference into the early development and use of cyanide leaching to recover gold. **8.5" X 11", 162 ppgs. Retail Price: $14.99**

California Gold Milling Practices - Unavailable since 1895 and released under the name "California Gold Practices", this rare publication provides an in depth look at early methods of milling used to reduce gold ores in California during the late 19th century. This volume provides a reference into the early development and use of milling equipment during the earliest years of the California Gold Rush up to the age of the Industrial Revolution. Much of the information still applies today and will be of use to small scale miners engaging in hardrock mining. **8.5" X 11", 104 ppgs. Retail Price: $10.99**

Leaching Gold and Silver Ores With The Plattner and Kiss Processes - Mining historian Kerby Jackson introduces us to a classic mining publication on the evaluation and examination of mines and prospects by C.H. Aaron. First published in 1881, it has been unavailable for over a century and sheds important light on the leaching of gold and silver ores with the Plattner and Kiss processes. **8.5" X 11", 204 ppgs. Retail Price: $15.99**

The Metallurgy of Lead and the Desilverization of Base Bullion - First published in 1896, it has been unavailable for over a century and sheds important light on the the recovery of silver from lead based ores. Some of the topics include the properties of lead and some of its compounds, lead ores such as galenite, anglesite, cerussite and others, the distribution of lead ores throughout the United States and the sampling and assaying of lead ores. Also covered is the metallurgical treatment of lead ores, as well as the desilverization of lead by the Pattinson Process and the Parkes Process. Hofman's text has long been considered one of the most important early works on the recovery of silver from lead based ores. **8.5" X 11", 452 ppgs. Retail Price: $29.99**

Ore Sampling For Small Scale Miners - First published in 1916, it has been unavailable for over a century and sheds important light on historic methods of ore sampling in hardrock mines. Topics include how to take correct ore samples and the conditions that affect sampling, such as their subdivision and uniformity. Particular detail is given to methods of hand sampling ore bodies by grab sample, pipe sample and coning, as well as sampling by mechanical methods. Also given are insights into the screening, drying and grinding processes to achieve the most consistent sample results and much more. **8.5" X 11", 124 ppgs. Retail Price: $12.99**

The Extraction of Silver, Copper and Tin from Ores - First published in 1896, it has been unavailable for over a century and sheds important light on how historic miners recovered silver, copper and tin from their mining operations. The book is split into three sections, including a discussion on the Lixiviation of Silver Ores, the mining and treatment of copper ores as practiced at Tharsis, Spain and the smelting of tin as it was practiced by metallurgists at Pulo Brani, Singapore. Also included is an overview and analysis of these historic metal recovery methods that will be of benefit to those interested in the extraction of silver, copper and tin from small mines. **8.5" X 11", 118 ppgs. Retail Price: $14.99**

The Roasting of Gold and Silver Ores - First published in 1880, it has been unavailable for over a century and sheds important light on how historic miners recovered gold and silver rom their mining operations. Topics include details on the most important silver and free milling gold ores, methods of desulphurization of ores, methods of deoxidation, the chlorination of ores, methods and details on roasting gold and silver ores, notes on furnaces and more. Also included are details on numerous methods of gold and silver recovery, including the Ottokar Hofman's Process, the Patera Process, Kiss Process, Augustin Process, Ziervogel Process and others. **8.5" X 11", 178 ppgs. Retail Price: $19.99**

The Examination of Mines and Prospects - First published in 1912, it has been unavailable for over a century and sheds important light on how to examine and evaluate hardrock mines, prospects and lode mining claims. Sections include Mining Examinations, Structural Geology, Structural Features of Ore Deposits, Primary Ores and their Distribution, Types of Primary Ore Deposits, Primary Ore Shoots, The Primary Alteration of Wall Rocks, Alterations by Surface Agencies, Residual Ores and their Distribution, Secondary Ores and Ore Shoots and Vein Outcrops. This hard to find information is a must for those who are interested in owning a mine or who already own a lode mining claim and wish to succeed at quartz mining. **8.5" X 11", 250 ppgs. Retail Price: $19.99**

www.ingramcontent.com/pod-product-compliance
Lightning Source LLC
Chambersburg PA
CBHW080600180526

45168CB00007B/2721